NORTH CAROLINA
STATE BOARD OF COMMUNITY COLLEGES
LIBRARIES
ASHEVILLE-BUNCOMBE TECHNICAL COMMUNITY COLLEGE

STP 985

Rapid Methods For Chemical Analysis of Hydraulic Cement

Ronald F. Gebhardt, editor

ASTM
1916 Race Street
Philadelphia, PA 19103

Library of Congress Cataloging-in-Publication Data

Rapid methods for chemical analysis of hydraulic cement.

(Special technical publication (STP) ; 985)
"ASTM publication code number (PCN) 04-985000-07."
Includes bibliographies and index.

1. Cement—Analysis. I. Gebhardt, Ronald F.
II. Series: ASTM special technical publication ; 985.
TP882.3.R36 1988   620.1'35   88-6302
ISBN 0-8031-0989-X

Copyright © by AMERICAN SOCIETY FOR TESTING AND MATERIALS 1988

NOTE

The Society is not responsible, as a body,
for the statements and opinions
advanced in this publication.

Peer Review Policy

Each paper published in this volume was evaluated by three peer reviewers. The authors addressed all of the reviewers' comments to the satisfaction of both the technical editor(s) and the ASTM Committee on Publications.

The quality of the papers in this publication reflects not only the obvious efforts of the authors and the technical editor(s), but also the work of these peer reviewers. The ASTM Committee on Publications acknowledges with appreciation their dedication and contribution of time and effort on behalf of ASTM.

Printed in Ann Arbor, MI
May 1988

Preface

In the late 1960s and early 1970s, a major effort was made by ASTM Subcommittee C1.23 on (then) Chemical Analysis to allow the use of rapid methods of analysis for cements. There was a basic consensus that this was needed because of the rapid shift away from classical analytical chemistry to instrumental methods of analysis. In fact, many laboratories, particularly those that were more progressive and had access to funds, had switched almost exclusively to rapid and instrumental methods for cement analysis. Many laboratories had not run a cement analysis in accord with ASTM Referee Methods for years and freely admitted that they no longer had the capability to do so. This had been a conscious decision, made largely because a full analysis by Referee Methods required a week.

Several instrumental methods were in fairly common use, such as atomic absorption spectrometry, X-ray (emission) spectrometry, and a spectrophotometric/titrimetric scheme developed by Z. T. Jugovic. Except for the Jugovic methods, there were several varieties of methods using any given instrument type, for example, AA or X-ray.

Consensus largely broke down at this point because of the tradition that analytical methods be printed in their entirety in the *Annual Book of ASTM Standards*. With at least three AA methods and three X-ray methods, all lengthy and apparently of equivalent (claimed) precision and accuracy, the Subcommittee faced a virtually insurmountable technical and political task. If one method of a generic type should be selected, the author would surely be gratified, but the authors of the competing methods would be offended, particularly since results using each method were approximately equivalent.

The approach of using a round-robin test program for all methods in an attempt to prove one better than the others was judged to be a more monumental task than desired by the subcommittee, even if enough participants could be found. Results from each method were likely to be about equivalent; the task would have taken several years; and it probably would have been inconclusive.

Someone, identity not known after these many years, suggested that the obvious approach for rapid and instrumental methods was to set standards for precision and accuracy that must be met by any method and then allow the analyst to select his own method so long as he could prove that it worked. This approach achieved consensus, resulting in the inclusion of Rapid Methods as Optional Methods in ASTM Methods for Chemical Analysis of Hydraulic Cement (C 114–77), which first appeared in the *1977 Annual Book of ASTM Standards*.

A list of references was (temporarily) appended to C 114, pending completion of this special technical publication (STP), to provide guidance to the analyst who might not have a rapid method readily at hand. At this same time, the Subcommittee promised to put together an STP which would contain in one place a number of examples of rapid methods for the analysis of hydraulic cement. That promise, after over ten years and several changes of editors through retirement, job changes, and other causes, is finally being met.

Even though there have been many improvements in instrumentation since most of these methods were written, the principles are still the same. Also, many analysts are using instruments very similar, if not identical, to those used in developing these methods since few of us have the inclination or funds to obtain every new model that comes out. In any event, the material provided herein provides guidance in the general field of instrumental analysis.

This STP is sponsored by ASTM Committee C-1 on Cement and Subcommittee C1.23 on Compositional Analysis.

Ronald F. Gebhardt
Lehigh Portland Cement Company,
Allentown, PA 18105, editor

Contents

Introduction   1

Section I. Atomic Absorption Methods

Method for Spectrochemical Analysis of Portland Cement Using an Atomic Absorption Spectrophotometer—PORTLAND CEMENT ASSOCIATION   5
APPENDIX: Evaluation of Method—E. H. SCOTT   12

Atomic Absorption Methods for Analysis of Portland Cement—E. H. SCOTT   15

Section II. X-Ray Spectrochemical Methods

Suggested Method for Spectrochemical Analysis of Portland Cement by Fusion with Lithium Tetraborate Using an X-Ray Spectrometer—CLYDE W. MOORE   31

Method for X-Ray Spectrochemical Analysis of Ground and Pelletized Cement or Clinker—J. E. MANDER AND C. W. TRADER   38

Suggested Method for X-Ray Emission Spectrometric Analysis of Portland Cement by the Energy-Dispersive Technique—B. D. WHEELER   48

Section III. Spectrophotometric/EDTA Methods

Spectrophotometric and EDTA Methods for Rapid Analysis of Hydraulic Cement—Z. T. JUGOVIC   57

Section IV. Free Lime Rapid Methods

The Determination of Free CaO in Cements and Clinkers—J. W. YULE AND R. D. CHADWICK   75

Extraction of Free Lime in Portland Cement and Clinker by Ethylene Glycol—M. P. JAVELLANA AND I. JAWED   78
APPENDIX: Evaluation of Method—N. T. FLORES   83

LaFarge Chemical Method No. 42: Free CaO—CANADA CEMENT LAFARGE LTD. 84

SECTION V. RAPID ANALYSIS OF SULFUR

Method for Analysis of Total Sulfur as SO_3 in Portland Cement and Clinker Using the LECO Sulfur Analyzer—M. G. LEWIS AND E. H. SCOTT 89

APPENDIX I— Alumina: Direct Determination Chemical Method 93

APPENDIX II— ASTM Methods for Chemical Analysis of Hydraulic Cement (C 114-85) 119

Introduction

The methods in this volume were submitted by their authors or organizations in response to a request for rapid and instrumental methods which were being used for analysis of hydraulic cements. This call for methods has been repeated almost annually since the mid-1970s. The result is the collection of methods contained herein that were submitted for publication. No claim is made as to completeness of coverage of rapid or instrumental methods, nor has any search of the literature been done. This is simply a compilation of the methods submitted.

The first three sections on Atomic Absorption, X-Ray, and Spectrophotometric Methods present procedures for typical "complete" analyses of cement. The last three sections contain methods for individual determinations. These methods are claimed by their authors or submitters to provide data in conformance with the requirements of ASTM Methods for Chemical Analysis of Hydraulic Cement (C 114) for Rapid Methods of Analysis (Appendix II). Some of the methods contain, or are followed by, data indicating such conformance. Except as contained herein, no attempt has been made to independently verify the precision, accuracy, or suitability of any method for its intended purpose.

Neither ASTM, Committee C-1 on Cement, Subcommittee C1.23 on Compositional Analysis, nor the Editor make any warranty, either express or implied, as to the suitability of any method contained herein for any purpose whatsoever. The material is presented only for information.

Caution: The methods contained herein indicate use of materials, operations, and equipment that are hazardous or potentially hazardous. This publication does not purport to address all, or even, necessarily, any of the aspects relating to safety. It is the responsibility of any user of the material contained herein to consult and establish appropriate safety and health precautions and practices and to determine the applicability of any regulatory limitations before use.

It is the intent of Subcommittee C01.23 on Compositional Analysis of Cement to be aware of advances in methods and techniques for analysis of hydraulic cements under the jurisdiction of ASTM Committee C-1 and for related material. The Subcommittee recognizes the desirability of dissemination of such information to the concerned analytical community. A publication such as this, a virtual handbook of analytical methods for cement, would seem to be a good medium for that purpose. The Subcommittee is contemplating future editions of this publication when warranted by sufficient new information. Consequently, since reliance must be placed on the practitioners, submission of material for future publication is solicited. Material need not be restricted to chemical analysis of cement. Methods for mineralogical analysis, identification of additions, and determination of constituents of blended cements will also be considered.

Ronald F. Gebhardt
Lehigh Portland Cement Company,
Allentown PA 18105; editor.

Section I
Atomic Absorption Methods

Two very similar methods are presented in this section. In fact, the second is based on the first with some modifications and comments that may be useful to some analysts.

They are based on a lithium metaborate fusion, which has essentially become the standard for analysis of cement and clinker. This approach also makes the method suitable for analysis of raw materials and raw meal.

Editor

Portland Cement Association[1]

Method for Spectrochemical Analysis of Portland Cement Using an Atomic Absorption Spectrophotometer

Introduction

This method is an update of that by Crow and Connolly of the Portland Cement Association published in the *Journal of Testing and Evaluation,* Vol. 1, No. 5, Sept. 1973, pp. 382–393. It was developed for a Perkin-Elmer double beam AA unit.

Although no data as to precision and accuracy were provided with the method, an evaluation was performed by E. H. Scott, of the Martin Marietta Cement Technical Center, (now defunct) in May of 1974. That evaluation is the appendix that immediately follows the method.

National Bureau of Standards (NBS) standard reference material cement samples are used for calibration (standardization?). Standards and samples are fused with lithium metaborate. The molten melt is transferred from the graphite fusion crucible to a beaker containing dilute nitric acid. Note that precious metal crucibles should never be used with borate fusions. After dissolution, the solution is filtered and diluted to volume with dilute nitric acid. This solution is used for analysis of silica (SiO_2), alumina (Al_2O_3), ferric oxide (Fe_2O_3), strontium oxide (SrO), manganese sesquioxide (Mn_2O_3), potassium oxide (K_2O), and sodium oxide (Na_2O). A more dilute solution containing lanthanum is used for CaO and MgO.

Editor

1. Scope

(a) This method describes the spectrochemical analysis of portland cement for the elements in the ranges indicated in Table 1.

(b) A series of solutions prepared from National Bureau of Standards (NBS) standard cement samples (or other carefully analyzed cements) is utilized for calibration purposes. The standard solutions chosen shall have concentrations that will bracket those expected in the solutions of the unknowns.

(c) An AA instrument equipped with a double beam optical system, signal averaging, and digital readout capability is necessary for the determination of calcium and silicon.

2. Apparatus and Equipment

(a) A double-beam atomic absorption spectrophotometer with a self-contained digital concentration readout, nitrous oxide-acetylene burner, and a three slot Boling air-acetylene

---

[1] This method was submitted by the Portland Cement Association in December 1972.

TABLE 1—*Ranges of elements used in the spectrochemical analysis.*

Element[a]	Concentration Range, %[b]
Calcium (CaO)	58 to 68
Silicon (SiO$_2$)	18 to 26
Aluminum (Al$_2$O$_3$)	0 to 7
Iron (Fe$_2$O$_3$)	0 to 6
Magnesium (MgO)	0 to 5
Potassium (K$_2$O)	0 to 1
Sodium (Na$_2$O)	0 to 1
Strontium (SrO)	0 to 0.5
Manganese (Mn$_2$O$_3$)	0 to 0.5

[a] The elemental analysis of portland cement is generally reported in terms of the oxide of the highest normal valence for each element.
[b] Results outside these ranges may be secured but will be of lesser accuracy unless special standards are available.

burner are suggested. Modifications of the following instructions may be necessary when otherwise-equipped instruments are used.

(b) *Mixing vessel*—A 15-mL platinum crucible or any noncontaminating vessel.

(c) *Graphite crucibles*,[3] 7.88-mL capacity, made from purified graphite—These should be preignited in a muffle furnace for 20 min at 950°C before use.

(d) *Clear plastic beakers* (polypropylene), 100-mL capacity.

(e) *Magnetic stirring bars*—The length of the bars used should be approximately $\frac{1}{2}$ in. (12.7 mm) less than the inside diameter of the plastic beakers.

(f) A source of clean moisture-free compressed air controlled by a low-pressure regulator—An air filter[4] with removable cartridge should be installed in the line.

(g) Cylinders of AA-grade acetylene and nitrous oxide with two-gage, two-stage pressure-reducing regulators.

(h) *Hollow cathode lamps*, of high spectral purity and high intensity.

(i) A 500-W voltage transformer for elimination of voltage fluctuations to the instrument.

(j) A vent to remove noxious fumes.

(k) *Muffle furnace*—The muffle furnace should be capable of continuous operation up to 1000°C and should have an indicating pyrometer accurate to ±25°C.

3. Reagents

(a) *Anhydrous lithium metaborate* (LiBO$_2$)—A blank determination should be made with each new supply of fusion material to determine possible contamination or background interferences.

(b) *Dilute nitric acid* (1 + 24)—Dilute 120 mL of concentrated nitric acid to 3 L with distilled water.

(c) *Lanthanum solution*—Add 200 mL of distilled water to 11.73 g of lanthanum oxide (La$_2$O$_3$) in a 600-mL beaker. While stirring, add 20 mL of hydrochloric acid. Warm the mixture and stir with a magnetic stirring bar until solution is complete. Cool to room temperature, filter into a 500-mL volumetric flask, dilute to the mark, and mix thoroughly. This solution contains 0.02 g of lanthanum per millilitre. A 99.99% pure reagent has been found satisfactory.

[3] Crucible 81574, from Ultra Carbon Co., Bay City, MI, has been found acceptable.
[4] Ultipore Filter from Perkin-Elmer Corporation has been found satisfactory.

(d) Standard samples—A supply of the National Bureau of Standards cement samples or other carefully analyzed cements.

4. Procedure

(a) Selection of standards—Select a series of at least four standard cement samples that will bracket the expected concentrations of the elements in the unknown samples.

(b) Preparation of standard and sample solutions—Grind a representative sample of cement or raw mix so as to pass a No. 100 (150-μm) sieve and mix thoroughly. Weigh 0.80 g of anhydrous lithium metaborate into a suitable mixing vessel, then add 0.5000-g cement. Mix thoroughly with a small Teflon® stirring rod. Transfer the mixture to a pre-ignited graphite crucible. Place the crucible in the mouth of an electric furnace and heat at 950°C for 5 min (or until the mixture melts). Remove from the furnace and gently swirl to coagulate any particles of fusion mix remaining on the walls of the crucible. Return to the furnace and heat for 10 min.

Remove from the furnace and immediately pour the molten melt into a clear polypropylene beaker containing 60 mL of nitric acid (1 + 24) and a Teflon®-coated magnetic stirring bar. Place the beaker on a magnetic stirring unit and stir for 10 min. Filter through a medium-textured filter paper into a 500-mL volumetric flask. Wash the beaker and filter paper thoroughly with nitric acid (1 + 24). Fill the flask to the calibration mark with dilute nitric acid (1 + 24), mix well, and transfer to a clean polypropylene bottle.

Prepare a fusion-blank solution in the same manner. A fresh fusion-blank solution should be prepared each time new batches of anhydrous lithium metaborate are used.

(c) Operating procedure for the determination of aluminum, iron, magnesium, manganese, potassium, sodium, and strontium oxides—Refer to Table 2 for general instrumental parameters, and to Table 3 for parameters that apply to certain instruments. Other instruments may require a different set of parameters:

(1) Prepare a series of sample and standard solutions as outlined in Sections 4(a) and (b).

(2) Dilute the solutions prepared in Section 4(c) (1), as shown in the method of sample preparation (Fig. 1). The final solution for calcium and magnesium should contain 10 mL of lanthanum solution.

(3) Adjust the monochromator to isolate the analytical line for the element to be determined, and peak the signal. (Adjustment of the slit width may be necessary on some instruments to control the purity and amount of radiation that reaches the photomultiplier.)

(4) Adjust the flow of gases to the burner and light the flame.

(5) Align the flame with the source beam by aspirating a solution of portland cement and adjusting the position of the burner so as to locate the optimum zone in the flame where maximum absorption occurs for the element of interest.

(6) Aspirate the fusion-blank solution and set the readout indicator to zero.

(7) Measure the absorbance of the standard and sample solutions sequentially until at least two readings are obtained for each of the solutions.

(8) Plot the average absorbance of the standards against the percent of the element of interest in the standards.

(9) Draw a working curve and determine the percentage of the element of interest in the unknown sample.

(10) If a digital readout system is used, results can be displayed directly as the percentage of the element being determined. To accomplish this, set the digital readout to read the

TABLE 2—General instrumental parameters.

Parameter	Aluminum	Calcium	Iron	Magnesium	Manganese	Potassium	Sodium	Silicon	Strontium
Wave length, Å	3092.6	4227	3719.9	2852.1	2794.8	7664.9	5890.0	2516.1	4607.3
Burner used	nitrous oxide	nitrous[a] oxide	nitrous[a] oxide	nitrous[a] oxide	Boiling	nitrous[a] oxide	nitrous[a] oxide	nitrous oxide	nitrous oxide
Concentration range, ppm	17 to 34	21 to 24	14 to 26	0 to 1.4	0 to 2	0 to 8	0 to 7	84 to 112	0 to 3.4
Flame	NO_2–C_2H_2 oxidizing fuel lean	air–C_2H_2 oxidizing fuel lean	air–C_2H_2 reducing (slightly fuel rich)	air–C_2H_2 reducing (fuel rich)	air–C_2H_2 reducing (fuel rich)	air–C_2H_2 reducing (slightly fuel rich)	air–C_2H_2 reducing (slightly fuel rich)	NO_2–C_2H_2 reducing (fuel rich)	NO_2–C_2H_2 oxidizing (fuel lean)

[a] A 2-in. (50.8-mm), short path, air acetylene burner can also be used.

ATOMIC ABSORPTION METHODS 9

TABLE 3—*Instrumental parameters for certain applicable atomic absorption spectrophotometers.*

Element	Wave Length, Å	Slit Setting	Lamp Current, mA	Burner Used	Air Pressure	Fuel Flow Reading Plastic Ball	Air Flow Reading Steel Ball	Concentration Range, ppm	Flame Used
Aluminum	3092.6	3	25	nitrous oxide	30	42	25	17 to 34	NO_2-C_2H_2
Calcium	4227	4	10	nitrous oxide	30	30	45	21 to 24	air-C_2H_2
Iron	3719.9	3	30	nitrous oxide	30	30	40	14 to 26	air-C_2H_2
Magnesium	2852.1	4	10	nitrous oxide	30	30	40	0 to 1.4	air-C_2H_2
Manganese	2794.8	4	25	Boling	30	40	60	0 to 2	air-C_2H_2
Potassium[a]	7664.9	4	12	nitrous oxide	30	30	40	0 to 8	air-C_2H_2
Sodium	5890.0	3	8	nitrous oxide	30	30	40	0 to 7	air-C_2H_2
Silicon	2516.1	3	35	nitrous oxide	30	47	27	84 to 112	NO_2-C_2H_2
Strontium	4607.3	3	25	nitrous oxide	30	47	30	0 to 3.4	NO_2-C_2H_2

[a] Spectral filter in light path, curvature correction used.

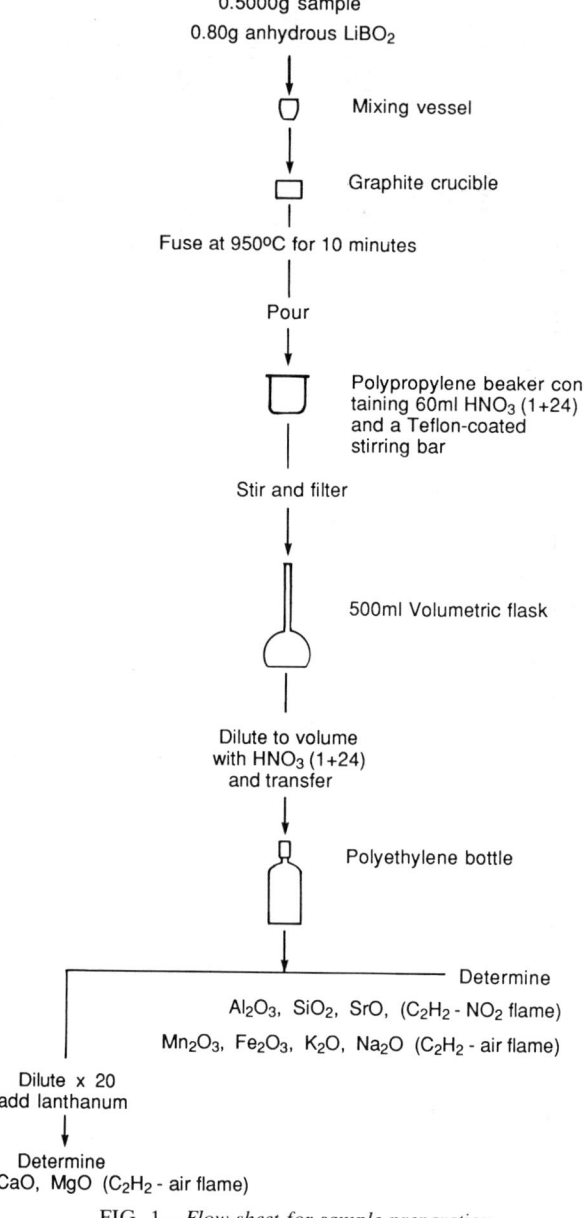

FIG. 1—*Flow sheet for sample preparation.*

proper percentage when aspirating a known standard solution, preferably the one near the high end of the concentration range. Standard solution can then be run alternately with sample solutions to check this use of the single-standard procedure.

(11) For nonlinear curves, "curve straightening" can be employed. This is done on certain instruments by using a curvature correction accessory.

(d) Operating procedure for the determination of silicon dioxide:

(1) Use the solution prepared in Section 4(c) (1).
(2) Adjust the instrumental parameters for silicon as shown in Table 2 and, for applicable instruments, as shown in Table 3.
(3) Aspirate a solution of portland cement and align the flame with the source beam by adjusting the position of the burner so as to locate the optimum zone in the flame where maximum absorption for silicon occurs.
(4) Vary the fuel adjustment to find the flame conditions for maximum absorption.
(5) Set the digital readout system so as to obtain the average of ten or more consecutive absorbance readings.
(6) Aspirate distilled water and set the readout indicator to zero.
(7) Measure the absorbance of the lowest cement standard, the sample, and the highest cement standard.
(8) Plot the absorbance of the standard solutions versus concentration (expressed as percent of silicon dioxide) and determine the approximate percentage of silicon dioxide in the sample.
(9) Adjust the digital readout for scale expansion. The standard solution having the highest silicon concentration should provide an absorbance of approximately 0.200 when aspirated into a nitrous oxide flame. Expand the value about eight times by adjusting the digital readout to read approximately 1400 (consult the instrument manual for digital readout device used).
(10) Set the digital readout system so as to obtain the average of at least 16 consecutive absorbance readings.
(11) Bracket the sample solution with standard solutions having silicon concentrations (expressed as silicon dioxide) approximately 2% above and below the sample's approximate concentration.
(12) Measure the expanded absorbance of the low standard, the sample, and the high standard sequentially until three values are obtained for each solution. Set the readout indicator to zero while aspirating a solution of distilled water after each series of three readings.
(13) Average these expanded absorbance values, plot the average value of the standards directly against percent silicon dioxide, and determine the percent silicon dioxide in the sample from the curve.

(e) Operating procedure for the determination of calcium oxide:

(1) Adjust the instrumental parameters as shown in Table 2, and for applicable instruments, as shown in Table 3.
(2) Aspirate the sample solution prepared for the determination of magnesium. Align the flame with the source beam by adjusting the position of the burner so as to locate the zone of the flame where maximum absorption for calcium occurs. Rotate the burner head to obtain an absorbance of approximately 0.375.
(3) Set the digital readout system so as to obtain the average of eight or more consecutive absorbance readings.
(4) Aspirate distilled water and set the readout indicator to zero.
(5) Measure the absorbance of the lowest cement standard, the sample, and the highest cement standard.
(6) Plot the absorbance of the standard solutions versus concentration (expressed as

percent of calcium oxide) and determine the approximate percentage of calcium oxide in the sample.

(7) Adjust the digital readout for scale expansion. Expand the absorbance value about seven times by adjusting the digital readout to display approximately 2600 (consult the instrument manual for digital readout device used).

(8) Set the digital readout system so as to obtain the average of 100 consecutive absorbance readings.

(9) Bracket the sample solution with standard solutions having calcium concentrations (expressed as calcium oxide) about 2% above and below the sample's approximate concentration. Set the readout device to obtain the average of at least 16 consecutive readings.

(10) Measure the expanded absorbance of the low standard, the sample, and the high standard sequentially until three values are obtained for each solution. Set the readout indicator to zero while aspirating a solution of distilled water after each series of three readings.

(11) Average these expanded values, plot the average values of the standards directly against percent calcium oxide (CaO), and determine the percent CaO from the curve.

5. Calculations

(a) To report the analyses of raw mix on the "as received basis" multiply the results obtained by 0.667.

APPENDIX

Evaluation of the Portland Cement Assocation Atomic Absorption Method by E. H. Scott[5]

Preface

This evaluation of the Portland Cement Association (PCA) Atomic Absorption Method was conducted by E. H. Scott, of the Martin Marietta Cement Technical Center, in May of 1974. It notes some precautions that may be necessary with the PCA method as written, while showing that the method can easily meet the requirements of ASTM Method for Chemical Analysis of Hydraulic Cement (C 114). Alkalies were not evaluated. It also notes some points that may be of value to improve instrument stability.

Editor

Introduction

An evaluation of the PCA procedure for the complete analysis of cement by AA has been carried out for use both by our laboratory and plant quality control applications. The procedure includes the determination of SiO_2, Al_2O_3, CaO, MgO, Fe_2O_3, SrO, and Mn_2O_3 in cement. The procedure is essentially as follows. A 0.5000-g sample of cement is fused with $LiBO_2$ in a carbon crucible for 15 min at 950°C. The hot melt is poured directly into dilute nitric acid (HNO_3). The fusion melt bead shatters and goes into solution after about

---

[5] Martin Marietta Cement Technical Center, Baltimore, MD 21227.

10-min stirring. The solution is filtered to remove carbon particles and diluted to 500 mL. SiO_2, Al_2O_3, Fe_2O_3, Mn_2O_3, and SrO are determined on this solution. For CaO and MgO analysis a 10-mL aliquot is diluted to 200 mL, and lanthanum is added to suppress chemical interferences.

Problems Encountered

No problems were encountered with Fe_2O_3, Al_2O_3, MgO, SrO, or Mn_2O_3. It was found necessary to employ curve corrections for Al_2O_3 and MgO to obtain the best precision and accuracy.

The only problem encountered with SiO_2 is salt buildup on the burner slit caused by the solids in solution and carbon buildup caused by the rich C_2H_2/N_2O flame required for the SiO_2 determination. It is necessary to thoroughly clean the burner slit with fine emery paper before SiO_2 analysis and to complete the analysis within approximately 20 to 30 min operating time to avoid erratic results. SiO_2 analysis requires care in setting up optimum instrument parameters to obtain good precision and accuracy.

CaO determination was a major stumbling block in the analysis of cement and still requires a great amount of care to obtain good accuracy and precision. The basic problem encountered was flame instability. CaO is extremely sensitive to flame instability. Flame instability can be the result of a multitude of factors. These include inadequate control of fuel and air flow, improper drainage, changes in back pressure, and changes in burner head temperature. It is also believed that the PCA practice of turning the burner head at a slight angle to reduce CaO sensitivity may contribute to instability caused by changes in gas flow along the burner slit.

A number of changes were made to improve stability for CaO analysis. The air compressor was reset to come on at 75 psi (517 kPa) and off at 90 psi (621 kPa). The regulator on the compressor was set at 60 psi (414 kPa), and the regulator on the instrument was set at 55 psi (379 kPa). This system thus provides a constant source of air flow to the instrument. The acetylene (C_2H_2) control knob on the instrument was wrapped with Teflon® tape to remove some of the play in the knob and provide much finer C_2H_2 control.

Three changes were made to improve drainage from the burner assembly. The assembly was dismantled and the smooth plastic lining apparently causes puddles of liquid to form and build up instead of allowing a continuous flow of liquid out of the burner head assembly. The mixing chamber was replaced with one of a new experimental design. The new mixing chamber has a sand blasted, heat treated surface, which should improve drainage properties.

TABLE 4—10×10 data.

Compound	Wet Analysis[a]		AA Analysis			
			Instrument		Method	
	Mean	σ	Mean	σ	Mean	σ
SiO_2	21.68	0.10	21.70	0.048	21.71	0.060
CaO	63.80[b]	0.08	63.76	0.071	63.76	0.088
Fe_2O_3	4.95	0.08	4.92	0.015	4.92	0.028
Al_2O_3	4.37[c]	0.12	4.35	0.016	4.36	0.020
MgO	2.45	0.07	2.53	0.013	2.53	0.017

[a] Average of 10 MM laboratories.
[b] Corrected for SrO and slight increase in loss on ignition (LOI).
[c] Corrected for TiO_2 and P_2O_5.

TABLE 5—*Qualification data.*

Compound	Difference Between Duplicates, Maximum (6 of 7 Required Within Limits)	Difference of the Average SRM Value (6 of 7 Required Within Limits)	AA Difference Between Duplicates	AA Difference of the Average SRM
SiO_2	0.16	±0.2	seven 0.0 to 0.10	seven 0.0 to 0.20
Al_2O_3	0.20	±0.2	seven 0.0 to 0.10	seven 0.0 to 0.10
Fe_2O_3	0.10	±0.10	seven 0.0 to 0.05	seven 0.0 to 0.04
CaO	0.20	±0.3	six 0.0 to 0.17 one 0.35	seven 0.0 to 0.20
MgO	0.16	±0.2	seven 0.0 to 0.11	seven 0.0 to 0.08
Mn_2O_3	0.03	±0.03	seven 0.0 to 0.01	seven 0.0 to 0.01
SrO	not established	not established	seven 0.0 to 0.01	seven 0.0 to 0.01

The plastic drain tube was shortened and plugged into the lower side of a 32 oz. (0.01 m^3) plastic bottle. A head of about 6 in. (152 mm) is maintained in both the bottle and in the plastic tube above the bottle. This modification also has the effect of maintaining an almost constant back pressure on the entire drainage system. The burner head is always allowed to warm-up for at least 20 min before CaO analysis to permit the head to reach thermal equilibrium. A greater aliquot dilution was made of the sample solution (10/500 instead of 10/200) to reduce sensitivity rather than turning the burner head at a slight angle.

Data

Once instrument parameters were established for CaO, SiO_2, Al_2O_3, Fe_2O_3, MgO, SrO, Mn_2O_3 and sample preparation procedures were established, data were run to establish accuracy and precision of the above constituents. A 10 × 10 was run on MM cement standards for CaO, SiO_2, MgO, Al_2O_3, and Fe_2O_3 against SRM cements to establish precision and accuracy of the method (Table 4). Ten portions of the same MM cement standard were weighed and prepared. Each portion was analyzed ten times. In addition, two sets of SRM cements 633, 634, 635, 636, 637, and 639 were prepared on nonconsecutive days and analyzed on nonconsecutive days (Table 5). This latter procedure was done to establish qualification data.

E. H. Scott[1]

Atomic Absorption Methods for Analysis of Portland Cement

Introduction

This method is very similar to the PCA method except for addition of titanium oxide (TiO_2) to the scheme of analysis, a more dilute solution for determination of calcium oxide (CaO) and magnesium oxide (MgO), and different operating conditions for CaO.

This method is somewhat more "complete" than the preceding method in that almost everything is step-by-step in the text. Potential problems and precautions are more fully discussed.

This method easily met the requirements of ASTM Method for Chemical Analysis of Hydraulic Cement (C 114). Note that lithium tetraborate can also be used in either method.

Editor

1. Scope

1.1 These methods cover the chemical analysis of portland[2] cements by atomic absorption for the oxides in the ranges indicated in Table 1.

The methods are based primarily upon methods developed by Portland Cement Association (PCA).[3] Major modifications have been made in the calcium oxide (CaO) analysis.[4] A cement sample is fused with lithium metaborate, and the fusion melt dissolved in dilute nitric acid. The solution is diluted to a desired volume. Cement standards are prepared in the same manner. The sample solution is analyzed against standard solutions. The solutions are aspirated into the flame of an atomic absorption spectrophotometer and the absorbance measured at the analytical absorption line of the desired element

$$\text{absorbance} = K \frac{1}{(100 - \%A)}$$

where

K = slope or scale expansion and
A = absorption.

---

[1] Martin Marietta Cement Technical Center, Baltimore, MD 21227.
[2] These methods are designed primarily for cements but could be expanded to include raw mixes as well.
[3] PCA Research and Development Bulletin RD027.01T, "Atomic Absorption Analysis of Portland Cement and Raw Mix Using Lithium Metaborate Fusion."
[4] The major difference in the CaO determination from that of the PCA method is a strontium internal standard is employed to compensate for flame instability. Also, a greater dilution (10 mL/500 mL) is made of the sample and standard solutions for the CaO and MgO determinations, and the burner head is kept parallel in the source beam for the CaO analysis instead of being turned at an angle.

TABLE 1—*Ranges of concentration for elements for chemical analysis.*

Element	Concentration Range, % Oxide
Calcium (CaO)	58 to 68
Silicon (SiO_2)	18 to 26
Aluminum (Al_2O_3)	0 to 7
Iron (Fe_2O_3)	0 to 6
Magnesium (MgO)	0 to 5
Potassium (K_2O)	0 to 1
Sodium (Na_2O)	0 to 1
Strontium (SrO)	0 to 0.5
Manganese (Mn_2O_3)	0 to 0.5
Titanium (TiO_2)	0 to 0.3

2. Methods

2.1 The methods appear in the order shown in Table 2.

3. Interference and Limitations

3.1 These methods were developed primarily for the analysis of portland cements. Limitations are noted in the procedure for specific constituents.

4. Apparatus and Materials

4.1 An 810 Jarrell-Ash atomic absorption spectrophotometer or equivalent with a self-contained digital readout, automatic zero, and curve correction. It is also equipped with a nitrous oxide burner head.

4.2 Mixing vessel—a 15-mL platinum crucible or any noncontaminating vessel.

4.3 Graphite crucibles, 7.88-mL capacity, made from purified graphite. Crucible, 815174, from Ultra Carbon Company, Bay City, MI, has been found acceptable. Crucibles should be preignited in a muffle furnace for at least 45 min at 950°C before use to remove the smooth crucible surface.

4.4 Plastic beakers, 250-mL capacity.

4.5 Magnetic stirrer and Teflon® coated stirring bars.

4.6 A source of clean moisture-free compressed air controlled by low-pressure regulators.

TABLE 2—*Order of methods.*

Method	Sections
Sample preparation	6
CaO	7 to 9
SiO_2	10 to 12
Fe_2O_3	13 to 14
MgO	13 to 14
Al_2O_3	13 to 14
Na_2O	13 to 14
K_2O	13 to 14
Mn_2O_3	13 to 14
SrO	13 to 14
TiO_2	13 to 14

An air filter with a removable cartridge should be installed in the line.

4.7 Pipette, 10 mL, 25 mL.

4.8 Cylinders of AA-grade acetylene and nitrous oxide with two-gage, two-stage pressure reducing regulators.

4.9 Hollow cathode lamps, of high spectral purity and high intensity.

4.10 A 500-W voltage transformer for elimination of voltage fluctuations to the instrument.

4.11 A vent to remove noxious fumes.

4.12 Muffle furnace—the muffle furnace should be capable of continuous operations up to 1000°C and should have an indicating pyrometer accurate to ±25°C.

5. Reagents

5.1 Anhydrous lithium metaborate ($LiBO_2$)—A blank determination should be made with each new supply of fusion material to determine possible contamination or background interferences for each element.

5.2 Dilute nitric acid (*1:25*)—dilute 120 mL of concentrated nitric acid to 3 L with distilled water.

5.3 Lanthanum solution—add 200 mL of distilled water to 11.73 g of lanthanum oxide (La_2O_3) in a 600 mL beaker. While stirring, add 20-mL of hydrochloric acid. Warm the mixture and stir with a magnetic stirring bar until solution is complete. Cool to room temperature, filter into a 500-mL volumetric flask, dilute to the mark, and mix thoroughly. This solution contains 0.02 g of lanthanum per millilitre.

5.4 Standards samples—A supply of National Bureau of Standards (NBS) cement samples or other carefully analyzed cements.

5.5 Strontium standard solution, 1000 ppm—weigh 1.684 g-strontium carbonate ($SrCO_3$) and dissolve in 20-mL $1M$ hydrochloric acid (HCl). Dilute to 1000 mL in a volumetric flask.

6. Preparation of Standard and Sample

6.1 Selection of standards—Select a series of four or five standard cement samples that will bracket the expected concentrations of the elements in the samples.

6.2 Preparation of Solution 1—This solution is for the determination of SiO_2, Al_2O_3, Fe_2O_3, MgO, Mn_2O_3, and SrO. Grind a representative sample of cement so as to pass a No. 100 (150-μm) sieve and mix thoroughly. Weigh 0.80 g of anhydrous lithium metaborate into a suitable mixing vessel, then add 0.5000-g cement. Mix thoroughly with a small Teflon® stirring rod or other noncontaminating stirring rod. Transfer the mixture to a pre-ignited graphite crucible. Place crucible in the mouth of an electric furnace and heat at 950°C for 5 min (or until the mixture melts). Remove from the furnace and gently swirl to coagulate any particles of fusion mix remaining on the walls of the crucible. Return to the furnace and heat for 10 min. Remove from the furnace, gently swirl to coagulate fusion mix, and immediately pour the molten bead into a plastic beaker containing 60 mL of dilute nitric acid (1:25) and a Teflon® coated magnetic stirring bar. Place the beaker on a magnetic stirring unit and stir for 10 min or until fusion bead is in solution. Filter through a medium textured filter paper, to remove small graphite particles, into a 500-mL volumetric flask. Wash the beaker and filter paper thoroughly with dilute nitric acid (1:25). Fill the flask to the calibration mark with dilute nitric acid (1:25), mix well, and transfer to a clean polyethylene bottle. Prepare a fusion-blank solution in the same manner. A fresh fusion-blank solution should be prepared each time new batches of anhydrous lithium metaborate are used.

6.3 *Preparation of Solution 2*—This solution is for the determination of CaO and MgO. Pipet 10 mL of each Solution 1 into 500-mL volumetric flasks. Into each flask pipet 25 mL of lanthanum solution and 10-mL strontium (1000-ppm) standard solution to bring strontium concentration to 20 ppm. Dilute to volume with demineralized water, and transfer to a clean polyethylene bottle. Prepare a reagent blank in the same manner including 25-mL lanthanum solution and 10-mL strontium standard solution.

6.4 *Preparation of Solution 3*—This solution is for the determination of TiO_2. Prepare a fusion solution as for Solution 1, but filter solution into a 100-mL volumetric flask. Make to volume with dilute nitric acid (1:25). Prepare a fusion blank in the same manner.

CaO

7. Scope

7.1 One of the primary variables that affects precision and accuracy of calcium analysis is flame instability. Compensation is made for this variable with the addition of a strontium internal standard. Lanthanum is added to mask the interferences of sulfate and phosphate in both CaO and MgO determinations. A stiff, lean, blue flame is the most satisfactory for calcium. The digital readout system, with its provisions for scale expansion and averaging, provides the optimum precision required for the calcium measurement. By expanding the relatively narrow range of absorbance values provided by the calcium solution, readability can be increased and precision improved. With signal averaging, these expanded absorbance readings can be integrated over a definite time interval, averaged, and displayed. Refer to Table 3 for general instrument parameters. Parameters should be optimized for each instrument.

8. Procedure

8.1 Burner slit should be completely clean inside and out before analysis.

8.2 Install calcium lamp in Channel A and strontium lamp in Channel B. Adjust monochromater in Channel A to isolate calcium line and adjust monochromater in Channel B to isolate Strontium line. Peak the signal in each channel.[5] Light flame and allow burner head 20 min to reach temperature equilibrium before analysis. Aspirate fusion blank and zero each channel. With instrument switched to Channel A and in absorbance mode optimize the instrument parameters while aspirating a high CaO portland cement standard. Align the flame with the source beam by adjusting the position of the burner so as to locate the optimum zone in the flame where maximum calcium absorption occurs. Recheck zero in each channel while aspirating fusion blank.

8.3 Switch instrument to A/B mode and absorbance mode for internal standardization. Aspirate reagent blank containing 20-ppm strontium and zero instrument in A/B mode. Switch instrument to concentration mode (instrument still in A/B mode). While aspirating high CaO standard expand concentration mode until nearly to full expansion, but do not allow digital readout to overfill. Rezero instrument while aspirating the reagent blank. Aspirate low standard and take five or six readings. Rezero instrument with reagent blank. Aspirate sample and take five or six readings. Rezero instrument with reagent blank. Aspirate high standard and take five or six readings. Obtain approximate level of sample by graphing average expanded scale readings versus percent for high and low standards.

---

[5] Carefully focus calcium lamp for maximum energy through-put. Focus strontium lamp so that the strontium beam will pass through same portion of flame as the calcium beam. This can be done by placing a white card at midpoint of burner head and perpendicular to light beam. The strontium beam should superimpose on the calcium beam.

TABLE 3—*Instrumentation*.

Element	Wave Length, Å	Slit, Å	Lamp Current, mA	Burner	Flame
Aluminum	3092.6	4	10	N_2O	N_2O/C_2H_2 reducing 8 mm red plume
Calcium	4227	2	10	N_2O	air/C_2H_2 oxidizing lean blue
Iron	3117.9	2	8	N_2O	air/C_2H_2 oxidizing lean, blue
Magnesium	2852.1	10	10	N_2O	air/C_2H_2 oxidizing lean, blue
Manganese	2794.8	4	10	N_2O	air/C_2H_2 oxidizing lean, blue
Potassium	7664.9	10	10	N_2O	air/C_2H_2 oxidizing lean, blue
Sodium	5890.0	4	8	N_2O	air/C_2H_2 oxidizing lean, blue
Silicon	2516.1	2	12	N_2O	N_2O/C_2H_2 reducing 1.5-cm red plume
Strontium	4607.3	10	12	N_2O	N_2O/C_2H_2 reducing 6-mm red plume

NOTE: Adjustment of the slit width may be necessary on some instruments to control the spectral purity and amount of radiation that reaches the multiplier. Oxidizer flow should be at maximum and full flow should be adjusted to obtain maximum absorbance for each element. Optimum burner head height should be established for each individual instrument and burner head for best results.

8.4 Rerun sample as before except bracket the sample with standards approximately 1 to 2% above and below sample. First sequence: low standard, sample, high standard. Second sequence; high standard, sample, low standard. Third sequence: low standard, sample, high standard. Take an average of the data. Obtain the level of CaO in the sample either by calculation or from a graph of average expanded scale readings versus percent CaO.

9. Calculation

Table 4 shows the calculation CaO.

SiO$_2$

10. Scope

10.1 Silicon is one of the less sensitive elements to be determined by atomic absorption, and all instrument parameters should be carefully optimized to obtain an absorbance in the

TABLE 4—*Calculation for CaO*.

Parameters	%CaO	Expanded Scale Readings			Average	Difference
High standard	65.5	1309	1304	1299	1304 ⎫	24
Unknown sample		1280	1283	1277	1280 ⎬	18
Low Standard	62.6	1260	1262	1264	1262 ⎭	—
ΔCaO	1.9					42

Unknown sample: %CaO = 62.6 + (1.9) (18/42) = 63.41

range of 0.2 to 0.5 to obtain the best precision. The digital readout system, with its provisions for scale expansion and signal averaging, provides the optimum precision required for the silicon measurement. By expanding the relatively narrow range of absorbance values provided by the silicon solution, readability can be increased and precision improved. With signal averaging, these expanded absorbance readings can be integrated over a definite time interval, averaged, and displayed. Refer to Table 3 for general instrument parameters. Parameters should be optimized for each instrument.

11. Procedure

11.1 Burner slit should be completely cleaned inside and out before analysis.

11.2 Install silicon lamp, and adjust monochromater to isolate the silicon line. Peak the signal. Carefully focus lamp for optimum energy through-put. Light flame and allow burner head 20 min for burner head to reach thermal equilibrium before analysis. Carefully optimize instrument parameters with instrument in absorbance mode. Aspirate a high SiO_2 portland cement standard and align the flame with the source beam by adjusting the position of the burner so as to locate the optimum zone in the flame where maximum absorption occurs. If flame appears to bounce or surge readjust nebulizer until condition no longer occurs. Vary the fuel adjustment to find the flame conditions for maximum absorption.

11.3 Switch instrument to concentration mode. Zero instrument while aspirating distilled water. Aspirate highest standard and expand concentration mode until nearly to full expansion, but do not allow digital readout to overfill. Rezero instrument while aspirating distilled water. Aspirate low standard and take five or six readings. Rezero instrument with distilled water. Aspirate sample and take five or six readings. Obtain approximate level of sample by graphing average expanded scale versus percent for high and low standards.

11.4 Rerun sample as before except bracket the sample with standards approximately 1 to 2% above and below sample. First sequence: low standard, sample, high standard. Second sequence: high standard, sample, low standard. Third sequence: low standard, sample, high standard. Take an average of the data. Obtain the level of SiO_2 in the sample either by calculation or from a graph of average expanded scale reading versus percent SiO_2.

12. Calculation

Table 5 shows the calculation for SiO_2.

MgO, Al_2O_3, Fe_2O_3, Na_2O, K_2O, SrO, Mn_2O_3, TiO_2

13. Scope

13.1 No serious difficulties are encountered with iron, magnesium, or manganese. Interferences associated with the determination of these elements are compensated for by the use of solutions prepared from standard cements.

TABLE 5—*Calculation for SiO_2.*

Parameters	SiO_2	Expanded Scale Readings			Average	Difference
High standard	21.9	1890	1900	1910	1900	100
Unknown sample	...	1795	1805	1800	1800	50
Low standard	20.7	1746	1750	1754	1750	
SiO_2	1.2					150

Unknown $\%SiO_2$ = 20.7 + (1.2) (50/150) = 21.10%

13.2 Sensitivities of sodium, potassium, and strontium would ordinarily be reduced by a high degree of ionization of these elements. But the presence of lithium in the solutions suppresses their ionization.

13.3 Critical adjustment of fuel/oxidant ratio is required to obtain maximum sensitivity for strontium and titanium.

13.4 The sensitivity of the aluminum determination is affected by several factors. These include the position of the burner relative to the path of the source beam, the fuel/oxidant ratio, and the efficiency of the nebulizer. The setting of these parameters to achieve optimum sensitivity is critical with the nitrous oxide flame. Very small departures from optimum settings can cause severe loss of sensitivity. Ionization of aluminum is suppressed by the presence of lithium, thereby increasing aluminum sensitivity. It is recommended that the aluminum beam be positioned 6 to 8 mm above the burner head.

13.5 Curve corrections may be necessary for the determination of Fe_2O_3, Al_2O_3, Na_2O, and K_2O depending upon the concentrations of these constituents and instrument sensitivity.

13.6 Refer to Table 3 for general instrument parameters for all elements. Parameters should be optimized for each instrument.

14. Procedure

14.1 Burner slit should be completely clean inside and out before analysis.

14.2 Install the lamp of the element to be determined and adjust the monochromater to isolate the desired line. Peak the signal. Focus lamp for maximum energy through-put. Adjust the flow of gases to the burner and light the flame. Align the flame with the source beam by aspirating a solution of portland cement and adjust the position of the burner so as to locate the optimum zone in the flame where maximum absorption occurs for the element of interest.

14.3 Aspirate the fusion-blank solution, and set the readout indicator to zero. Measure the absorbance of the standard solutions and sample solution sequentially until at least three readings are obtained for each of the solutions. Repeat and average readings.

14.4 Plot the average absorbance of the standards against the percentage concentration of the element of interest in the standards. Draw a working curve and determine the percentage of the oxide determined in the sample.

14.5 If the working curves show that absorbances are nearly linear with concentrations, and the absorbance concentration curve passes through the origin, satisfactory results can be obtained by employing a single high standard for each of these oxides. The instrument

TABLE 6—*Precisions for the constitutents in cement.*

Oxide	σ
CaO	0.071
SiO_2	0.060
Al_2O_3	0.020
Fe_2O_3	0.028
MgO	0.017
Na_2O	0.015
K_2O	0.015
SrO	0.004
Mn_2O_3	0.005
TiO_2	0.005

TABLE 7—Qualification data.

NBS SRM Number	NBS Value	Run #1	Run #2	Difference Between Runs	Specific Difference Allowed Between	Average of Two Runs	Difference of Average From SRM	Specific Difference Allowed
				Al_2O_3				
633	3.79	3.74	3.75	0.01	0.20	3.74	0.05	±0.2
634	5.20	5.13	5.18	0.05	...	5.15	0.05	
635	6.20	6.17	6.20	0.03	...	6.18	0.02	
636	3.06	2.98	3.06	0.08	...	3.02	0.04	
637	3.30	3.30	3.22	0.08	...	3.26	0.04	
638	4.46	4.45	4.45	0.00	...	4.45	0.01	
639	4.30	4.27	4.32	0.05	...	4.29	0.01	
				CaO				
633	64.47	64.40	64.40	0.00	0.20	64.40	0.07	±0.3
634	62.60	62.52	62.55	0.03	...	62.53	0.07	
635	59.80	59.77	59.80	0.03	...	59.78	0.02	
636	63.49	63.48	63.44	0.04	...	63.46	0.03	
637	66.00	66.00	66.05	0.05	...	66.02	0.02	
638	62.10	62.00	62.20	0.20	...	62.10	0.00	
639	65.80	65.80	65.85	0.05	...	65.82	0.02	
				SiO_2				
633	21.88	21.84	21.80	0.04	0.16	21.82	0.06	±0.2
634	20.73	20.75	20.81	0.06	...	21.78	0.05	
635	18.41	18.45	18.45	0.00	...	18.45	0.04	
636	23.22	23.25	23.15	0.10	...	23.20	0.02	
637	23.07	23.10	23.10	0.00	...	23.10	0.03	
638	21.48	21.42	21.40	0.02	...	21.41	0.07	
639	21.61	21.50	21.65	0.15	...	21.57	0.04	
				Mn_2O_3				
633	0.04	0.04	0.04	0.00	0.03	0.04	0.00	±0.03
634	0.28	0.28	0.28	0.00	...	0.28	0.00	
635	0.09	0.09	0.10	0.01	...	0.09	0.00	
636	0.12	0.11	0.11	0.00	...	0.11	0.01	
637	0.06	0.06	0.06	0.00	...	0.06	0.00	

ATOMIC ABSORPTION METHODS

Sample								Tolerance
638	0.05	0.04	0.04	0.00	0.04	...	0.01	
639	0.08	0.08	0.08	0.00	0.08	...	0.00	

MgO

Sample								Tolerance
633	1.04	1.05	1.04	0.01	1.04	...	0.00	±0.2
634	3.33	3.41	3.40	0.01	3.40	0.16	0.07	
635	1.25	1.25	1.24	0.01	1.24	...	0.01	
636	3.98	4.03	4.01	0.02	4.02	...	0.04	
637	0.68	0.71	0.71	0.00	0.71	...	0.03	
638	3.86	3.92	3.90	0.02	3.91	...	0.05	
639	1.28	1.29	1.29	0.00	1.29	...	0.01	

TiO$_2$

Sample								Tolerance
633	0.24	0.24	0.25	0.01	0.24	...	0.00	±0.03
634	0.30	0.29	0.31	0.02	0.30	0.03	0.00	
635	0.32	0.30	0.32	0.02	0.31	...	0.01	
636	0.17	0.17	0.16	0.01	0.16	...	0.01	
637	0.21	0.21	0.21	0.00	0.21	...	0.00	
638	0.25	0.25	0.23	0.02	0.24	...	0.02	
639	0.31	0.31	0.30	0.01	0.30	...	0.01	

K$_2$O

Sample								Tolerance
633	0.165	0.167	0.170	0.003	0.168	...	0.003	±0.05
634	0.43	0.41	0.42	0.01	0.41	0.03	0.02	
635	0.45	0.42	0.44	0.02	0.43	...	0.02	
636	0.57	0.55	0.57	0.02	0.56	...	0.01	
637	0.245	0.242	0.255	0.013	0.248	...	0.003	
638	0.59	0.57	0.58	0.01	0.57	...	0.02	
639	0.05	0.05	0.06	0.01	0.05	...	0.00	

Na$_2$O

Sample								Tolerance
633	0.64	0.65	0.64	0.01	0.64	...	0.00	±0.05
634	0.14	0.16	0.15	0.01	0.15	0.03	0.01	
635	0.07	0.08	0.08	0.00	0.08	...	0.01	
636	0.10	0.11	0.11	0.00	0.11	...	0.01	
637	0.13	0.15	0.15	0.00	0.15	...	0.02	
638	0.12	0.14	0.13	0.01	0.13	...	0.01	
639	0.65	0.65	0.64	0.01	0.64	...	0.01	

SrO

Sample								Tolerance
633	0.31	0.31	0.30	0.01	0.30	no spec	0.01	no spec
634	0.12	0.12	0.12	0.00	0.12	...	0.00	

Table 7—(Continued)

NBS SRM Number	NBS value	Run #1	Run #2	Difference Between Runs	Specific Difference Allowed Between	Average of Two Runs	Difference of Average From SRM	Specific Difference Allowed
635	0.21	0.21	0.21	0.00	...	0.21	0.00	
636	0.04	0.04	0.04	0.00	...	0.04	0.00	
637	0.09	0.09	0.10	0.01	...	0.09	0.00	
638	0.07	0.07	0.07	0.00	...	0.07	0.00	
639	0.15	0.15	0.14	0.01	...	0.14	0.10	
Fe_2O_3								
633	4.20	4.19	4.20	0.01	0.10	4.19	0.01	±0.10
634	2.85	2.90	2.88	0.02	...	2.89	0.04	
635	2.62	2.59	2.57	0.02	...	2.58	0.04	
636	1.61	1.63	1.63	0.00	...	1.63	0.02	
637	1.80	1.82	1.80	0.02	...	1.81	0.01	
638	3.58	3.60	3.61	0.01	...	3.60	0.02	
639	2.40	2.42	2.46	0.04	...	2.44	0.02	

can be calibrated in the concentration mode with the high standard to readout directly in percent.

14.6 For nonlinear curves, "curve straightening" can be employed by using curvature correction accessory.

15. Precision of Methods

The precisions shown in Table 6 have been attained for the constituents in cement.

Precisions were obtained by preparing ten fusion samples of one portland cement and determining each of the oxides ten times for each of the fusion samples.

16. Qualification Data.

Table 7 shows qualification data.

SECTION II
X-Ray Spectrochemical Methods

Three methods are presented in this Section. Two are wavelength dispersive (standard) methods wherein the elements are detected and measured separately; this may be either sequential or simultaneous depending on the instrument. The third is an energy dispersive method in which all elements are detected and measured simultaneously with the same detector.

The first method analyzes a flux-fused sample and is typical of the type of procedure in which the fused sample is then ground and pelletized. The second method uses a pressed powder sample without fusion. This particular method uses no additives for grinding aid or binder that remain in the sample after preparation. Sample preparation is, thus, much simplified. The third method uses a typical pressed powder sample preparation in which the sample and binder/grinding aid must be weighed.

All of the methods use interelement corrections as has been found necessary for general analysis of cement and clinker. Interelement corrections can be avoided in some cases if analysis is restricted to samples from a single source and if the composition range is sufficiently narrow; that is not the case when qualifying under the requirements of ASTM Methods for Chemical Analysis of Hydraulic Cement (C 114).

Particle size effects are not significant in any of the methods included. The fusion eliminates it in the first method, the particle size requirement eliminates it in the second, and the third makes provision for it by testing with X-rays.

The mineralogical effect, variations due to the crystalline species in which the various elements appear, is not considered in any of the methods. When analysis is restricted to portland cement and clinker, mineralogy is so similar that this effect is not significant. It also is not significant, obviously, in any fused sample so long as the original constituents dissolve in the melt. When using pressed powder samples, the analyst should be aware that the mineralogical effect may be significant for some raw materials, raw meals, and blended cements.

Editor

Suggested Method for Spectrochemical Analysis of Portland Cement By Fusion with Lithium Tetraborate Using an X-Ray

Introduction

This method is comparable to several fusion methods that have been published from time to time. Note that the various automatic fusion devices are suitable for use with this method with adjustment, if necessary, of quantities of materials. Some methods use lithium metaborate in place of tetraborate.

A similar method was developed in the early 1960s by I. Adler and H. Rose, of the U.S. Geological Survey, Washington, DC. This was a "heavy absorber" method to eliminate any need to make interelement corrections and allow the use of smaller samples. The formula was 100 mg of sample, 100 mg of lanthanum oxide, and 500 mg of lithium tetraborate. This was fused into a bead, ground, and spread carefully over the surface of a 1-in. (25-mm) backing of boric acid, which had been prepared with a flat surface. For X-ray units that take larger samples, quantities would have to be increased. The method was developed for rocks and minerals where only small amounts might be available. The heavy absorber limits analysis to the specimen surface so large amounts of material are not required, but the entire surface must be covered since any void will cause low results. Count rates are lowered by both the heavy absorber and the greater dilution. If freedom from interelement effects is a major consideration and the instrument to be used is sufficiently stable, this approach would have some promise.

Fused flat specimens can also be used without grinding. Care must be taken to assure that the specimens are uniform and without segregations. Required quantities of materials are larger, but the problem of delamination of the specimen from the backing is eliminated. Since the fusion product, whether ground or not, is either stable or only slowly reactive, specimens are essentially permanent and can be reused over long periods of time.

Editor

Appears in *Methods for Analytical Atomic Spectroscopy*, 8th ed., E2 SM 10-26, American Society for Testing and Materials, 1987, pp. 949–955.

Suggested Method for
SPECTROCHEMICAL ANALYSIS OF PORTLAND CEMENT BY FUSION WITH LITHIUM TETRABORATE USING AN X-RAY SPECTROMETER[1]

SUBMITTED BY CLYDE W. MOORE[2]

1. Scope

1.1 This method covers the X-ray spectrochemical analysis of portland cement for eight constituents in the concentration ranges indicated:

Constituent	Concentration Range, percent
Alumina (Al_2O_3)	3.0 to 7.0
Calcium oxide (CaO)	62.0 to 68.0
Ferric oxide (Fe_2O_3)	1.5 to 5.5
Magnesium oxide (MgO)	0.4 to 5.0
Potassium oxide (K_2O)	0.02 to 1.00
Silica (SiO_2)	19.0 to 25.0
Strontium oxide (SrO)	0.05 to 0.40
Titanium dioxide (TiO_2)	0.10 to 0.40

NOTE 1—This method depends on the determination of the elements, but the convention of expressing composition in terms of oxides is followed.

2. Summary of Method

2.1 The sample of portland cement is mixed with lithium tetraborate to provide a flux to sample ratio of 3:1 in the fused bead. The fused bead is allowed to cool and is ground to a fine powder from which a briquet is prepared. The flat surface of the briquet is irradiated by X rays. The secondary X rays produced are dispersed by means of analyzing crystals and the intensities are measured by detectors at selected wave lengths. Pulses of the correct amplitude from the flow-proportional counters and the scintillation counters are selected by single-channel pulse-height selectors and counted by electronic counting circuits for a pre-determined time interval. A suitable permanent reference standard is irradiated and measured in a like manner.

2.2 Intensity ratios are obtained by dividing the X-ray intensity from a given element in the sample by the X-ray intensity from the same element in the reference standard. Concentrations in terms of the oxide are determined from analytical curves, or from algebraic equations relating the X-ray intensity ratio and the loss of ignition to concentration.

3. Definitions

3.1 For definitions of terms used in this method, refer to ASTM Definitions E 135, Terms and Symbols Relating to Emission Spectroscopy.[3]

3.2 *Reference standard*—as used in this method, is a permanent reference pellet that will emit secondary X-rays of stable intensities in day-to-day operation when placed under the excitation conditions used on unknown samples. The intensities should be in the range encountered with portland cements that have been prepared as described in this method.

4. Apparatus

4.1 *Sample Preparation Equipment:*

4.1.1 *Glass Bottle*, 30-ml capacity with wide mouth.

4.1.2 *Fusion Crucibles*, graphite may be obtained commercially or may be cut from a 32-mm diameter graphite rod. The dimensions

---

[1] This suggested method has no official status in the Society, but is published as information only. The method is based on the experience of the submitter. Comments are solicited.
[2] Lone Star Cement Corp., Technical Department, P.O. Box 2148, Houston, Tex. 77001.
[3] Appears in this publication.

shall be 29 mm high over-all, with a cavity of 9-ml capacity and a slightly tapered interior bottom.[4]

4.1.3 *Muffle Furnace*, capable of continuous operation at 1000 C with an indicating pyrometer accurate to ±25 C.

4.1.4 *Grinder*—A rotary swing mill with tungsten carbide grinding chamber and a time control is suitable.[5]

4.1.5 *Briquetting Press*, to provide a controllable pressure of at least 7 kgf/mm² (10,000 psi), and produce a pressed pellet suitable in size for X-ray analysis.

4.2 *Excitation Source*, capable of operation at 50 kV and 50 mA constant potential with input voltage regulated to ±1 percent. The X-ray tube shall have a target of chromium, or other material of equal or better efficiency for exciting elements of atomic number 12 to 14, and a beryllium window.

4.3 *Spectrometer Equipment*,[6] providing for determination of selected X-ray intensities in the wavelength range 0.877 to 9.889 Å from a sample at an air pressure of 250 μm or less, controllable to ±50 μm.

4.3.1 *Crystals and Collimators*, chosen to provide satisfactory count rates over background for determinations at each specific wavelength. Lithium fluoride, ethylenediamine d-tartrate, pentaerythritol, or ammonium dihydrogen phosphate crystals have proven satisfactory (Table 1).

4.3.2 *Detectors*—Two types shall be used, gas flow proportional counters and scintillation counters, both of which shall provide linear count rate response at the count rates encountered. The scintillation counter shall be used in the determination of strontium, and it is preferable for the determination of iron.

4.4 *Measuring System*, consisting of detectors, preamplifiers, linear amplifiers, pulse height selectors, electronic pulse counting circuits, and a means of displaying the number of counts collected during the measuring cycle.

5. Reagents

5.1 *Detector Gas*, consisting of a mixture of 90 percent argon and 10 percent methane.

5.2 *Lithium Tetraborate* (anhydrous $Li_2B_4O_7$ powder) reagent grade or 4-9s plus purity, for use as a flux.

5.3 *Boric Acid*, technical grade, for use as a filler or backing material for structural strength of the pressed pellet.

6. Standards

6.1 National Bureau of Standards portland cement samples identified as Standard Reference Materials (SRM) Nos. 1011, 1013, 1014, 1015, and 1016, or their replacements, shall be used as standard samples for preparing analytical curves.

6.2 A reference as defined in 3.2.

7. Preparation of Standards and Samples

7.1 Consult the certificate of analysis in order to calculate the weight fraction loss on ignition (L_s) of the standards. When vials are freshly opened, the certificate value for the loss may be used. After opened vials have been stored, the loss should be redetermined as described below for samples. Determine the loss on ignition (L_s) of samples as specified in Section 36 of ASTM Method C 114, for Chemical Analysis of Hydraulic Cement.[7] The weight loss during the initial heating period may be used in this procedure. Record the decrease in weight in grams to the nearest 0.0001 g. This will equal the weight fraction lost on ignition.

7.2 Obtain the loss on ignition (L_f) for the lithium tetraborate as follows: (Note 2) Weigh 1 g of $Li_2B_4O_7$ in a tared porcelain crucible recording the weight to the nearest 0.0001 g. Cover and ignite the crucible and its contents in a muffle furnace at a temperature of 1000 C for 10 min. Record the decrease in weight in grams. This equals the weight fraction lost on ignition.

NOTE 2—It is normally sufficient to check the loss on ignition when a new container is opened. After long use or after storage, the loss on ignition should be redetermined.

7.3 Weigh a quantity of sample that will yield 0.7000 g upon fusion. Calculate quantity as follows:

---

[4] Suitable crucibles may be obtained from Spex Industries, Metuchen, N. J. 08840.
[5] A suitable grinder may be obtained from Spex Industries, Metuchen, N. J. 08840, or from Applied Research Laboratories, Inc., Sunland, Calif. 91040.
[6] The Philips Electronic Instruments, Mount Vernon, N. Y. 10550, Model PW-1250 X-ray Spectrometer or equivalent is suitable.
[7] *Annual ASTM Book of Standards*, Part 9.

Weight of sample = $0.7000/(1.0000 - L_s)$

Weigh a quantity of $Li_2B_4O_7$ that will yield 2.1000 g upon fusion. Determine quantity as follows:

Weight of flux = $2.1000/(1.0000 - L_f)$

7.4 Mix the portland cement and the flux in a glass bottle by shaking and rolling. A mixer mill may be used but is not required. Transfer the mixture quantitatively to the fusion crucible. Heat the crucible in the muffle furnace at 1000 C for 10 min. Allow the melt to cool in the crucible and transfer the cooled bead to the grinder and grind for 1 min (Note 3). Place a quantity of inert material in the press mold cavity (Note 4). Spread the ground fused powder on top of the inert backing material.[8] Press to the selected pressure and hold for 30 s (Note 5). Release the pressure and label the briquet.

NOTE 3—For grinders other than the rotary swing mill, the grinding time may be determined from a study of X-ray intensities for different grinding times on identical pellets. The correct grinding time is the minimum time required to give stable intensities.

NOTE 4—Boric acid and cellulose powder have made good backing materials.

NOTE 5—Higher pressures normally improve X-ray intensities obtainable from the lighter elements. Once a pressure is selected it should be standardized for both the standards and the samples. Pressure in the range from 14 kgf/mm² to 35 kgf/mm² (20,000 psi to 50,000 psi) have proven most suitable.

7.5 Prepare a suitable permanent reference pellet which can be irrdadiated in the same manner as the standards and samples. Fuse a selected portland cement (SRM No. 1014 is suitable) with $Li_2B_4O_7$ in the 3:1 ratio. Permit the melt to cool slowly on a flat graphite surface in a mold of suitable size for the X-ray spectrometer. Grind and polish a flat surface on the fused glass-like mass.

7.6 Prepare synthetic samples containing only $Li_2B_4O_7$ and the oxide of the element to be determined. For the intended purpose, these synthetic samples may be either fused or intermixed. The amount of oxide should be adjusted to give count rates near the maximum encountered in portland cement samples.

8. Preparation of Apparatus

8.1 Within the limitations of 4.3, select suitable analyzing crystals, collimators, and detectors for the determination of each individual oxide. Following the instrument manufacturer's instructions, use the synthetic samples to make the following adjustments: (a) set the detector gas flow rate and permit the detectors to be adequately flushed with the detector gas, (b) set the analyzing crystals to give maximum count rates at the specified wavelength with the appropriate synthetic samples in the sample holder, (c) establish the optimum detector voltage, and (d) set the pulse height selectors with the appropriate synthetic sample in place (Note 6). Table 1 shows the analytical lines to be used, suggested selections for detectors and analyzing crystals, and the minimum number of counts that should be accumulated for SRM No. 10.14.

NOTE 6—In the determination of MgO in portland cement, the phosphorus in the ADP crystal fluoresces, and the phosphorus radiation reaches the detector along with the magnesium radiation. Calcium in the sample is primarily responsible for this fluorescence. The pulse height selector can adequately separate the magnesium radiation if precautions are taken. Determine the pulse amplitude distribution curve with a sample in place containing only MgO in $Li_2B_4O_7$ (0.04 g of MgO in 2.67 g $Li_2B_4O_7$ fused and ground as in 7.3). Repeat the procedure with a sample in place containing only CaO in $Li_2B_4O_7$ (0.46 g of CaO in 2.34 g of $Li_2B_4O_7$ fused and ground). Repeat the procedure with a sample in place containing both MgO and CaO (0.04 g of MgO, 0.46 g of CaO in 2.30 g of $Li_2B_4O_7$ fused and ground). From these three pulse amplitude distribution curves it is possible to identify the pulses caused by the Mg K-α radiation and to set the limits of the pulse height selector to obtain a maximum of the magnesium pulses and a minimum of the phosphorus pulses.

9. Excitation and Radiation Measurements

9.1 *Excitation and Exposure*—Select an exposure time and power setting that will give at least the number of counts indicated in Table 1 when the permanent reference standard is in place. Introduce the sample into the sample chamber, taking care not to contaminate the surface that is to be exposed to the X-ray beam. Evacuate the chamber, and turn on the sample spinner if one is available. Use the same excitation conditions and exposure time for all standards, samples, and for

---

[8] Aluminum crucibles are available from Spex Industries, Metuchen, N. J .08840, and from Philips Electronic Instruments, Mount Vernon, N. Y. 10550, which may be used in certain mold assemblies where added structural strength of the pressed briquet is needed.

the reference standard. Measure the reference standard immediately before or after each sample.

9.2 *Radiation Measurements*—Obtain the radiation measurements of the analytical line for the element i in the sample pellet (I_{si}) and in the reference standard (I_{ri}). Form the intensity ratio for each element to be determined:

$$R_i = I_{si}/I_{ri}$$

10. Calibration and Standardization

10.1 *Analytical Curves*—Prepare and run the five Standard Reference Material portland cements three times on each of three separate days. Plot the concentration on the ignited basis on the ordinate and the X-ray intensity ratio on the abscissa for each constituent. See Appendix A1 for methods of determining the equation of the analytical curve.

10.2 Acceptable accuracy can generally be obtained for all constituents within the concentration ranges specified without taking interelement effects into account. However, the accuracy for Fe_2O_3, MgO, and SiO_2 is usually improved by considering interelement effects. A method for applying the corrections is described in Appendix A1.

11. Calculations

11.1 Obtain the X-ray intensity ratios. Read the concentration on the ignited basis from the analytical curves, and multiply the results by $(1 - L_s)$. Alternatively, obtain the intensity ratios and calculate the oxide analysis directly from the equations relating intensity ratio and loss on ignition to the oxide content of the cement.

12. Precision and Accuracy

12.1 *Precision*—The single-operator-day precision and the single-operator multi-day precision have been determined.

12.1.1 *Single-Operator-Day Precision*—Ten portions of a cement sample were prepared, and the X-ray data were obtained in a single day. The precision was calculated as follows:

$$s = \sqrt{\Sigma d^2/(n - 1)}$$

where:
s = standard deviation,
d = difference between individual determinations and the mean of the ten determinations, and

n = number of samples.

This standard deviation represents the single-operator-day precision.

12.1.2 *Single-Operator Multi-Day Precision*—One portion of the same cement sample was prepared on each of six separate days and the X-ray intensities were measured. The precision was calculated from the above equation.

12.1.3 *Coefficients of Variation* were calculated as follows:

$$v = 100\,s/\bar{x}$$

where:
v = coefficient of variation of the method,
s = standard deviation, and
$\bar{x}$ = average concentration in percent.

The results of the precision determinations are given in Table 2.

12.2 *Accuracy*—The distribution of the National Bureau of Standards SRM sample data points around the analytical curves was used to estimate accuracy. The standard deviation was calculated as follows:

$$s = \sqrt{\Sigma(X - x)^2/(n - 1)}$$

where:
s = standard deviation,
X = certificate value of the concentration on the ignited basis,
x = value of concentration on the ignited basis calculated from the least squares equation, and
n = number of determinations.

Fifteen determinations were made representing the five standards run three times each on three separate days. No data points were discarded. The accuracy shown in Table 3 represents the accuracy obtained from single determinations.

DISCUSSION

The accuracy of this fusion method has been demonstrated to be superior to the accuracy attainable by briquetting techniques not involving fusion. The method is particularly useful where it is desired to adopt a single procedure and single analytical curves to cover all portland cements. Three of the minor constituents of portland cement, K_2O, TiO_2, and SrO, can be determined on the same sample with a high degree of precision and accuracy. Very little additional time is required to determine these components.

An additional advantage of the method is that it relates analyses directly to Standard Reference Materials through the analytical curves. Accuracy of the curves may be checked by analyzing one of the NBS cements. New curves may be established when the need arises.

Maintaining a constant flux to sample ratio in the fused bead is important where the loss on ignition varies from near zero to the specification limits, or beyond, because X-ray intensity ratio output from the elements is a nonlinear function of the percent flux in the fused bead. In single-plant operations where the loss on ignition is relatively uniform, an average value for the loss can probably be used in daily analyses to determine the weight of sample to use. The 3:1 ratio tends to stabilize the mass absorption coefficient while maintaining convenient X-ray intensity levels. This tends to reduce interelement interference.

This suggested method applies specifically to portland cement. However, the sample preparation procedure is directly applicable to the analysis of many of the raw materials used in the manufacture of portland cement. Appropriate standard samples and slightly modified interelement corrections are required. The analysis of limestone, shell, clay, shale, sand, and some coal ash can be performed with an appropriate set of analytical curves.

TABLE 1 Analytical Lines and Detector Arrays

Element	Analytical Line, Å	Analyzing Crystal[a]	Detector	Minimum Counts[b] for Reference Standard
Al	$K\alpha$ 8.338	PET	gas flow	3.1×10^3
Ca	$K\beta$ 3.090	EDDT	gas flow	1.9×10^6
Fe	$K\alpha$ 1.937	LiF	scintillation	4.8×10^6
K	$K\alpha$ 3.742	EDDT	gas flow	1.0×10^4
Mg	$K\alpha$ 9.889	ADP	gas flow	6.0×10^3
Si	$K\alpha$ 7.126	PET	gas flow	1.3×10^6
Sr	$K\alpha$ 0.877	LiF	scintillation	2.0×10^3
Ti	$K\alpha$ 2.750	PET	gas flow	2.5×10^3

[a] PET = pentaerythritol, EDDT = ethylenediamine d-tartrate, ADP = ammonium dihydrogen phosphate, and LiF = lithium fluoride. This selection of analyzing crystals has proven satisfactory but is not necessarily optimum in terms of best count rate over background. Other choices within limitations of 4.3.1 and 4.3.2 may be made.

[b] This is the minimum number of counts on SRM No. 1014 to achieve the precision listed in Table 2.

TABLE 2 Data on Precision[b]

Oxide	Average Concentration, percent	Single-Operator-Day Precision (1S), percent	Single-Operator Multi-Day Precision (1S), percent	Single-Operator Multi-Day Coefficient of Variation
Al_2O_3	3.79	0.07	0.05	1.32
CaO	66.71	0.09	0.14	0.21
Fe_2O_3,[a]	2.65	0.007	0.006	0.23
K_2O	0.40	0.005	0.005	1.25
MgO	1.26	0.04	0.05	3.97
SiO_2	22.05	0.05	0.05	0.23
TiO_2	0.17	0.005	0.005	2.94

[a] Fe_2O_3 data corrected for calcium interelement effect.

[b] Data for Table 2 and Table 3 supplied through the courtesy of Cecil E. Carter, Research Chemist, Lone Star Cement Corp.

TABLE 3 Data on Accuracy for Single Determinations[b]

Oxide	Average Concentration, percent	Standard Deviation, percent	Coefficient of Variation
Al_2O_3	5.0	0.18	3.60
CaO	65.0	0.30	0.46
Fe_2O_3,[a]	3.5	0.02	0.57
K_2O	0.5	0.01	2.00
MgO	2.7	0.05	1.85
SiO_2	22.0	0.12	0.55
SrO	0.23	0.01	4.35
TiO_2	0.25	0.01	4.00

[a] Standard deviation on Fe_2O_3 without correction for interelement effect of calcium is 0.04.

[b] Data for Table 2 and Table 3 supplied through the courtesy of Cecil E. Carter, Research Chemist, Lone Star Cement Corp.

APPENDIXES

A1. Equations of the Analytical Curves

A1.1 By the mathematical method of least squares, determine the equation of the best straight line relating the concentration of the oxides of element i on the ignited basis (c_i) to the X-ray intensity ratio (R_i) for element i.

$$c_i = mR_i + b$$

where m and b are constants of the equation which are determined by the method of least squares. The concentration of the oxide of element i in the original cement (C_i) then becomes:

$$C_i = (mR_i + b)(1 - L_s).$$

A2. Interelement Corrections

A2.1 Accuracy is usually improved in the determination of MgO, Fe_2O_3, and SiO_2 by applying interelement corrections. Methods that have been found helpful are as follows:

A2.1.1 *MgO*—The correction to MgO is required when the combination of detectors and pulse height selectors cannot adequately differentiate between the energy level of the magnesium radiation from the sample and the energy level of the phosphorus radiation from the ADP analyzing crystal. The calcium in the sample is largely responsible for causing the phosphorus to fluoresce. The following linear correction has proven useful:

$$c_{Mg} = aR_{Mg} + dR_{Ca} + f$$

where:

c_{Mg} = concentration of MgO in the sample on the ignited basis,
R_{Mg} = X-ray intensity ratio for magnesium,
R_{Ca} = X-ray intensity ratio for calcium, and
a, d, and f = constants to be determined from data on standard samples.

The concentration of MgO in standard samples is known. The corresponding X-ray intensity ratios for Mg and Ca are determined by the above suggested method. From the data, the constants a, d, and f are determined by the mathematical method of least squares. For convenience in making analytical curves the equation can be rearranged to give:

$$c_{Mg} = a(R_{Mg} + d/a\,R_{Ca}) + f$$

and the quantity in parentheses may be considered a corrected intensity ratio, R'_{Mg}.

$$R'_{Mg} = R_{Mg} + d/a\,R_{Ca}$$

Then the concentration of MgO is a linear function of the corrected intensity ratio:

$$c_{Mg} = aR'_{Mg} + f$$

A2.1.2 Fe_2O_3—Correction for the interelement effect of calcium on iron is, in effect, a correction for change in the mass absorption coefficient of the fused pellet. Therefore the relationship is exponential. A corrected X-ray intensity ratio is calculated for Fe as follows:

$$R'_{Fe} = R_{Fe}\exp gR_{Ca}$$

where:

R'_{Fe} = corrected X-ray intensity ratio for Fe,
R_{Fe} = uncorrected X-ray intensity ratio for Fe,
R_{Ca} = X-ray intensity ratio for calcium,
e = base of natural logarithms, 2.718, and
g = constant to be determined from data on standard samples.

The concentration of Fe_2O_3 is a linear function of R'_{Fe} and may be expressed by the equation:

$$c_{Fe} = \gamma R'_{Fe} + \beta$$

c_{Fe} = concentration of the ferric oxide on the ignited basis, and
γ, and β = constants to be determined from data on standard samples.

In practice it is convenient to assume a value for the constant g and proceed to calculate γ and β by the method of least squares using data from the standards. Determine the sum of the squares of the deviations as follows:

$$E^2 = \Sigma(X - x)^2$$

where:

E^2 = sum of the squares of the deviations,
X = certificate value for concentration of Fe_2O_3 on the ignited basis, and
x = calculated value for concentration of Fe_2O_3 using the constants g, γ, and β.

Repeat the procedure with successive approximations of the constant g until an acceptable minimum value for E^2 is found. Three or four approximations for the constant g will usually lead to an acceptable equation. Under conditions outlined in the suggested method the constant g is a decimal fraction approximately equal to 0.6.

A2.1.3 SiO_2—Correction for interelement effects on silicon is similar in nature to that described above for iron. In portland cements the interferring elements, in decreasing order of importance, are aluminum, magnesium, and iron. If corrections for all three are to be made the following expression can be used to calculate a corrected intensity ratio.

$$R'_{Si} = R_{Si}\exp(\lambda R_{Al} + \delta R'_{Fe} + \epsilon R'_{Mg})$$

where:

R_{Al} = intensity ratio for aluminum, and
λ, δ, and ϵ = constants to be determined from standards.

The other terms are defined as previously indicated. The concentration of SiO_2 on the ignited basis is a linear function of the corrected intensity ratio of silicon.

$$c_{Si} = m_1 R'_{Si} + b_1$$

where m_1 and b_1 are constants to be determined from data on standards. When corrections are to be made for all three interferring elements there are five constants to be evaluated. Therefore, more than five standards are needed in order to provide statistical reliability in the constants. If only one correction is to be made, the procedure described above for ferric oxide may be followed.

A2.1.4 *CaO*—Interelement effects on the determination of CaO have not been clearly discernible to the submitter under the conditions described in this method. However, the mass absorption coefficient of potassium for the calcium K radiation is high. Therefore a correction may be desirable if the potassium content exceeds the limits outlined in 1.1. If a correction is desired, it can be applied by a procedure analogous to that outlined in A2.1.2.

J. E. Mander[1] *and C. W. Trader*[1]

Method for X-Ray Spectrochemical Analysis of Ground and Pellitized Cement or Clinker

This method uses the as-is sample ground to high fineness and briquetted. The sample, in the preferred method, is ground with Freon TF as a grinding aid. All of the Freon is evaporated after grinding, so no time-consuming weighings are required. Consequently, it is even possible to use "volumetric proportioning" if desired, so long as quantities of material from sample to sample are not so different as to significantly affect grinding. Experience has shown that when nonvolatile grinding aids or binders are used, materials must be proportioned exactly by weighing on an analytical balance to achieve satisfactory results.

A significant drawback with this method for cement and clinker can be the poor physical stability of the specimens. While they can be kept in a desiccator for several days to weeks, even then the specimens begin to hydrate and expand. Useful specimen life outside a desiccator is several hours to a few days. Normally this is no problem, especially for quality control or certification analysis, since the sample need only be retained until completion and initial verification of the analysis. If longer retention is required as for secondary standards, disputed analyses, or field problems, a well-homogenized large sample of the material can be retained in a sealed container and fresh specimens prepared as needed. This approach has also been used to provide periodic checks (daily or by shift) and to provide quality control of sample preparation and instrument stability.

An example of ASTM Method for Chemical Analysis of Hydraulic Cement (C 114) qualification data using this method with a Philips PW 1270 X-ray analyzer is attached. The method has also been used to qualify two ARL 74 000 X-ray analyzers.

Editor

1. Scope

1.1 This paper provides a method for X-ray spectrochemical analysis of portland cement and clinker using selected interelemental corrections.

1.2 The method is applicable to the determination of up to eleven constituents in the concentration ranges indicated in Table 1.

NOTE 1—This method is based on elemental analysis, but the convention of expressing composition in terms of oxides is followed.

NOTE 2—The concentration ranges are extrapolated from the composition ranges of National Bureau of Standard (NBS) samples 633, 634, 635, 636, 637, 638, and 639. The method is applicable to concentrations outside the range, provided other suitable standardization samples are used.

---

[1] Martin Marietta Cement Technical Center, 1450 S. Rolling Rd., Baltimore, MD 21227.

X-RAY SPECTROCHEMICAL METHODS   39

TABLE 1—*Concentration ranges of constituents.*

Constituent	Concentration Range
Silica, SiO_2	18.0 to 25.0
Aluminum oxide, Al_2O_3	3.0 to 7.0
Ferric oxide, Fe_2O_3	1.5 to 5.5
Calcium oxide, CaO	58.0 to 67.0
Magnesium oxide, MgO	0.5 to 5.5
Sulfur trioxide, SO_3	1.0 to 7.0
Potassium oxide, K_2O	0.02 to 1.0
Sodium oxide, Na_2O	0.04 to 1.0
Titanium dioxide, TiO_2	0.05 to 0.60
Strontium oxide, SrO	0.05 to 0.60
Phosphorus pentoxide, P_2O_5	0.05 to 0.50

2. Summary of Method

2.1 A sample of portland cement or clinker is ground in a rotary grinding mill, briquetted, and irradiated as soon as possible after preparation. The characteristic radiation from each element in the sample is dispersed by Bragg diffraction from selected crystals, and X rays of selected wave lengths are measured by detectors sensitive to those wave lengths. The detector output is related to concentration of each element. Detector outputs (intensities) are inserted into a computer programmed to calculate each oxide analysis from calibration curves and then to correct selected elements for interelemental effects.

3. Definition of Terms

For definition of terms used in this method, refer to ASTM Terminology Relating to Emission Spectroscopy (E 135).

4. Apparatus

4.1 *Sample Preparation Equipment, Reagents, and Materials*

4.1.1 *Automatic Grinding Mill* capable of reducing the samples to a maximum size fraction of 14 μm. The mill shall have a wearing surface of tungsten carbide or other material, which will not cause detectable contamination of the sample.[2]

4.1.2 *Briquetting Press* with controls to provide for reproducible pressure and capable of compacting at pressures of 700 kg/cm^2 (10 000 psi), or greater.

4.1.3 *Balance and Weights* conforming to 4.1 of ASTM (C 114).

4.1.4 *Aluminum Sample Cups* or other suitable material for sample backing.

4.1.5 *1,1,2-trichlorotrifluoroethane*,[3] or other suitable grinding aids.

4.1.6 *Infrared Drying Lamp.*

4.1.7 *Computer.*

4.2 *Excitation Source* rhodium or chromium X-ray tube powered by a constant-potential X-ray generator. The input voltage to the power supply must be regulated.

4.3 *Measuring System*—Spectrometer providing for determination of selected X-ray intensities. One channel per element, specifically designed to detect and measure the element of interest; each channel consists of suitable collimators, crystals, and detectors.

---

[2] A suitable automatic grinding mill can be purchased from Spex Industries, Inc., Metuchen, NJ, or from Applied Research Laboratories, Inc., Sunland, CA.

[3] Du Pont "Feron TF" works well.

5. Standards

National Bureau of Standard portland cements identified as Standard Reference Materials (SRM) Numbers 633, 634, 635, 636, 637, 638, and 639 may be used as standard samples for preparing analytical curves. Other materials with accurately known oxide concentrations may be used. All values should be true oxide compositions.

6. Preparation of X-Ray Apparatus

6.1 Follow the manufacturer's instructions for initial assembly, conditioning, and preparation of the X-ray fluorescence apparatus.
6.2 Follow the manufacturer's instructions with respect to control settings.
6.3 Once the settings are established, they should be adhered to as closely as possible.

7. Preparation of Sample

7.1 Grind 5.0 g of material with 15 mL of Freon for at least 3 min (when standardizing on SRM cements, use the entire vial of cement). The grinding time necessary to reduce the sample to -15 μm may vary with the age and condition of the grinding mill. Tests should be conducted to determine the proper grinding time.

NOTE 3—Other suitable grinding aids may be used provided both the sample and the grinding aid have been analytically weighed to a fixed proportion. Grinding aids that are completely volatilized before briquetting the sample need not be weighed.

7.2 Remove grinding container from mill and place under infrared drying lamp. A fan that slowly moves air across the surface of the sample will shorten drying time.
7.3 Carefully brush the dried sample onto glazed paper and homogenize the sample.
7.4 Transfer sample into an aluminum sample cup.[4]
7.5 Briquet sample at a compacting pressure of 700 kg/cm^2 (10 000 psi) or greater.

8. Excitation and Radiation Measurements

8.1 *Excitation and Exposure*—Introduce the sample into the analyzing chamber of the X-ray instrument as quickly as possible after its preparation taking care not to scar or contaminate the surface of the sample. Produce and record intensities from all elements whose concentrations are to be determined.

8.2 *Measurement Statistics*—Continue to collect X-ray intensities until the statistical variations have been reduced to a level that will allow the analysis to meet the specifications on difference between duplicates shown in Table 1 of ASTM C 114.

9. Calibration

9.1 *Preparation and Analysis*—Prepare the seven NBS Standard Reference Material portland cements according to Sections 7.1 to 7.5 and analyze as described in Sections 8.1 and 8.2.

---

[4] A quantity of inert material is frequently used as a backing material instead of aluminum cups. "Boraxo," boric acid, and cellulose powder have been found to work satisfactorily.

TABLE 2—*Interelement correction α factors for cement.*

Analyte Interfering Element	CaO		SiO$_2$		Al$_2$O$_3$	
	Rh	Cr	Rh	Cr	Rh	Cr
Na$_2$O	−0.0013	−0.0007	+0.0075	+0.0110	+0.0079	+0.0096
MgO	−0.0018	−0.0013	+0.0065	+0.0100	+0.0110	+0.0120
Al$_2$O$_3$	−0.00053	+0.0002	+0.0098	+0.0130	...	...
SiO$_2$	+0.0003	+0.0013	...	...	−0.0037	−0.0050
SO$_3$	+0.0020	+0.0030	+0.0027	+0.0030	+0.0078	+0.0020
K$_2$O	+0.0228	+0.0240	−0.0001	+0.0019	+0.0020	+0.0049
P$_2$O$_5$	−0.00016	−0.0015	−0.0003	+0.0008	−0.0003	−0.0007
CaO	...	...	+0.0010	−0.0015	−0.0007	+0.0043
Fe$_2$O$_3$	−0.0028	−0.0019	+0.0071	+0.0086	+0.0093	+0.0088
LOI[a]	−0.0067	−0.0060	−0.0013	−0.0013	−0.0006	+0.0028

[a] Loss on ignition.

9.2 Analytical Curves—Uncorrected

9.2.1 For all elements (except silica (SiO$_2$), alumina (Al$_2$O$_3$), and calcium oxide (CaO)) to which interelemental corrections are not to be applied, plot the oxide concentrations in each standard sample versus its X-ray intensity, or intensity ratio; or by mathematical means determine the equation for the line relating the concentration of each oxide to its X-ray intensity ratio.

9.2.2 Compare the differences between NBS values and the values determined from the equations or plots in Section 9.2.1 to the permissible variations in Table 1 of ASTM C 114 in order to verify acceptable accuracy.

9.3 Analytical Curves—Corrected for Interelemental Effects

9.3.1 Determine the apparent concentrations for SiO$_2$, Al$_2$O$_3$, and CaO by applying corrections for interelemental absorption and enhancement effects (see Appendix A2 for description of the calculations).

9.3.2 For SiO$_2$, Al$_2$O$_3$, and CaO, plot the apparent concentrations versus their X-ray intensities, or intensity ratios; or by mathematical means determine the equation for the line relating the apparent concentrations to the X-ray intensities or intensity ratios.

9.3.3 Compare the differences between the apparent concentrations for SiO$_2$, Al$_2$O$_3$, and CaO with the values determined from the equations or plots for an initial check of validity of the calibration.

9.3.4 Using the values for SiO$_2$, Al$_2$O$_3$, and CaO determined from the equations or plots in 9.3.2 and the values for all other elements determined in Section 9.2.1, apply the interelemental corrections to get "calculated actual concentrations" for SiO$_2$, Al$_2$O$_3$, and CaO. Compare the differences between these "calculated actual concentrations" and the NBS values to Table 1 of ASTM, C 114 in order to verify acceptable accuracy (see Appendix A3 for a description of the calculations).

10. Analysis of Unknown Samples

10.1 *Preparation and Analysis*—Prepare the unknown sample according to Sections 7.1 and 7.5 and analyze as described in Sections 8.1 and 8.2.

10.2 *Calculation of Concentrations using Interelemental Corrections*—Obtain "actual concentrations" of unknown samples using the method described in Appendix A3.

APPENDICES

A1. Instrumental Stability

A1.1 *Instrumental Drift*—Once the instrument has been calibrated, it is necessary to check continually each element being analyzed to insure that the original calibration is being maintained. Logs of these data should be initiated and reviewed periodically to identify any drift trend.

A1.2 *"Drift" Standards*—Corrections for instrumental drift can be made by using two previously constructed "drift" standards. Intensities are obtained on these standards immediately after calibration, and oxide values determined by the calibration equations are assigned to them. The assigned oxide percentages are compared with X-ray intensities; new slopes and intercepts are calculated when necessary. For SiO_2, Al_2O_3, and CaO the changes are made with respect to the "apparent concentrations." There are limits beyond which the "drift standards" will not adequately compensate for instrumental drift. If there are extreme variances in X-ray intensities recalibration may be necessary.

A1.2.1 Autoclaved cement briquets or fused glass pellets have been used as "drift" standards. In order for the autoclaved briquets to represent sufficient differences on the calibration curve, it is suggested that one of these standards be diluted with lead oxide, approximately 50% by weight, to insure that this standard represents the lower portion of the calibration curves at all times. Other types of "drift" standards may be used as long as they remain essentially unchanged over long periods of time.

A2. Interelemental Corrections

The interelemental corrections are taken from the paper "X-Ray Fluorescence Analysis of Portland Cement Through The Use of Experimentally Determined Correction Factors."[5] The correction factors from this article are listed in Table 2.

EXAMPLE: To calculate an apparent concentration, use the analysis for NBS SRM 633 from Table 3, and the α factors from Table 2 for a chromium target X-ray tube, and the following equation

$$\%\text{apparent} = \%\text{actual}/[1 + \Sigma\,(\alpha \text{ factor}) \times (\%\text{interfering element})]$$

Then the apparent concentration for SiO_2 is calculated as shown in Table 4. Table 5 contains the apparent concentrations for SiO_2, Al_2O_3, and CaO for all seven SRMs.

TABLE 3—*NBS SRM 633 analysis.*

%SiO_2 = 21.88	%SO_3	= 2.20	TiO_2	= 0.24[a]
%Al_2O_3 = 3.79	%K_2O	= 0.165	Mn_2O_3	= 0.04[a]
%Fe_2O_3 = 4.2	%Na_2O	= 0.64	SrO	= 0.31[a]
%CaO = 64.47	%P_2O_5	= 0.24	ZnO	= 0.01[a]
%MgO = 1.04	%LOI	= 0.75	B	= 0.01[a]
			F	= 0.08[a]

[a] Not used in interelemental corrections.

[5] Anderson, C. H., Mander, J. E., and Leitner, J. W., *Advances in X-ray Analysis*, Vol. 17, pp. 214–224.

TABLE 4—*Calculation of apparent concentration for SiO_2.*

Interfering Element	(α Factor)		% Interfering Element		(Corrections)
Al_2O_3	(+0.0130)	×	(3.79)	=	(+0.0492700)
Fe_2O_3	(+0.0086)	×	(4.2)	=	(+0.0361200)
CaO	(−0.0015)	×	(64.47)	=	(−0.0967050)
MgO	(+0.0100)	×	(1.04)	=	(+0.0104000)
SO_3	(+0.0030)	×	(2.20)	=	(+0.0066000)
K_2O	(+0.0019)	×	(0.165)	=	(+0.0003135)
Na_2O	(+0.0110)	×	(0.64)	=	(+0.0070400)
P_2O_5	(+0.0008)	×	(0.24)	=	(+0.0001920)
LOI	(−0.0013)	×	(0.75)	=	(−0.0009750)
			Σ corrections	=	+0.01222555

%Apparent SiO_2 = 2.188/(1 + Σ corrections) = 21.88/1.01222555 = 21.615

A3. Calculating Concentrations with Interelemental Corrections

The actual concentrations of SiO_2, Al_2O_3, and CaO can be estimated by successive applications of the correction equation using the following procedure.

Calculate the apparent concentrations for SiO_2, Al_2O_3, and CaO from the X-ray intensities as in Section 9.3.2, and calculate the actual concentrations for all other elements. Using these concentrations (%apparent for SiO_2, Al_2O_3, CaO, and %actual for all others), calculate the correction factors for SiO_2, Al_2O_3, and CaO. The correction factors are then applied to the apparent concentrations to get intermediate corrected concentrations for SiO_2, Al_2O_3, and CaO. New correction factors are calculated by using the corrected concentrations for SiO_2, Al_2O_3, and CaO and the actual concentrations for all other elements. The new correction factors are applied to the original *apparent concentrations,* and intermediate corrected concentrations are again obtained. This process is repeated until the differences between the new intermediate corrected concentrations and last calculated corrected concentrations are insignificantly small. When the differences are small enough the last calculated corrected

TABLE 5—*Apparent concentration.*

NBS SRM	%SiO_2	%Al_2O_3	%CaO
	Chromium Tube		
633	21.62	3.08	62.89
634	19.96	4.18	61.06
635	17.62	5.04	58.01
636	23.08	2.49	61.28
637	23.70	2.74	64.15
638	20.60	3.57	60.16
639	21.52	3.51	64.21
	Rhodium Tube		
633	19.02	4.00	64.91
634	17.83	5.41	63.06
635	15.74	6.31	60.09
636	20.40	3.23	63.28
637	20.54	3.62	66.15
638	18.44	4.59	62.18
639	18.86	4.60	66.22

Note: Values are calculated from NBS (Interim Summary Report) Feb. 1975.

concentrations become the calculated actual concentrations. The sequence of calculations is summarized as follows:

Correction Equation

$$\%\text{corrected} = (\%\text{apparent}) \times [1 + \Sigma\,(\alpha\text{ factor}) \times (\%\text{conc})]$$

where

%conc = %actual for all but SiO_2, Al_2O_3, and CaO,
%apparent for SiO_2, Al_2O_3, and CaO the 1st time through loop,
%previous corrected for SiO_2, Al_2O_3, and CaO each time through loop except first.

EXAMPLE: To calculate the actual concentrations for NBS SRM 633, use the apparent concentrations for NBS SRM 633 from Table 5, for a chromium tube and the above equation where SiO_2 = 21.615, Al_2O_3 = 3.080, and CaO = 62.890 treating them as if the concentrations were derived from the equations in 9.2.1 and 9.3.2 and using interelemental corrections from Table 2 for chromium tube.

1st Loop Through Calculations

A. %corrected SiO_2 = (21.615) × [1 + Σ (α factor) × (%conc)]

where

1 + Σ (αFactor) × (%conc) =
1 + (0.013 × (%Al_2O_3 = 3.080) + (0.0086) × (%Fe_2O_3 = 4.20) +
(−0.0015) × (%CaO = 62.890) + (0.01) × (%MgO = 1.04 +
(0.003) × (%SO_3 = 2.20) + (0.0019) × (%K_2O = 0.165) +
(0.011) × (%Na_2O = 0.64) + (−0.0013) × (%LOI = 0.75) +
(0.0008) × (%P_2O_5 = 0.24)

= 1.00539550

∴ = %corrected SiO_2 = 21.615 × 1.00539550 = 21.73162373

B. %corrected Al_2O_3 = [3.080 × (1 + Σ (α factor) × (%conc)]

where

1 + Σ (α factor) × (%conc) =
1 + (−0.005) × (%SiO_2 = 21.615) + (0.0088) × (%Fe_2O_3 = 4.20) +
(0.0043) × (%CaO = 62.890) + (0.012) × (%MgO = 1.04) +
(0.002) × (SO_3 = 2.20) + (0.0049) × (%K_2O = 0.165) +
(0.0096) × (%Na_2O = 0.64) + (0.0028) × (%LOI = 0.75) +
(−0.0007) × (%P_2O_5 = 0.24)

= 1.22449338

∴ = %corrected Al_2O_3 = 3.080 × 1.22449338 = 3.77143962

TABLE 6—*Differences between Loops 1 and 2.*

Loop	SiO_2	Al_2O_3	CaO
First	21.875	3.790	64.470
Second	21.732	3.771	64.458
	0.143	0.019	0.012
	$\Sigma = 0.174$		

C. Following the above procedure

$$\%\text{corrected CaO} = 62.890 \times 1.02492540 = 64.4575833$$

2nd Loop Through Calculations

Use $SiO_2 = 21.73162373$, $Al_2O_3 = 3.77143961$, and CaO = 64.4575833 when calculating $1 + \Sigma$ (α factor) $\times$ (%conc)

$$\text{new }\%\text{corrected SiO}_2 = 21.615 \times 1.01203288 = 21.87509065$$

$$\text{new }\%\text{corrected Al}_2\text{O}_3 = 3.080 \times 1.23051655 = 3.78999097$$

$$\text{new }\%\text{corrected CaO} = 62.890 \times 1.02511562 = 64.46952109$$

The differences between the last corrected concentrations (from Loop 2) and the corrected concentrations from Loop 1 are too great; another loop through the calculations is necessary. The sum of the differences should be less than 0.01.
The differences between loops 1 and 2 are shown in Table 6.

3rd Loop Through Calculations

$$\text{new }\%\text{corrected SiO}_2 = 21.615 \times 1.01225610 = 21.877991562$$
$$\text{new }\%\text{corrected Al}_2\text{O}_3 = 3.080 \times 1.23054386 = 3.79007510$$
$$\text{new }\%\text{corrected CaO} = 62.890 \times 1.02512191 = 64.46991663$$

The differences between the corrected concentrations from Loop 3 and the corrected concentrations from Loop 2 are small enough; therefore the "calculated actual concentrations" are equal to the corrected concentrations from Loop 3.
The above calculations are easily handled with a simple iteration by a computer.

A4. Qualification Data

TABLE 7—*X-ray qualification data, 1977. Instrument: PW 1270 with Cr tube.*

SRM	Certified Value	% Oxide 1st Run	% Oixde 2nd Run	Average % Oxide	Difference of Duplication	Difference of Average From SRM
		SiO_2 (Corrected for Si and Ca ΔM)				
633	21.90	21.91	21.85	21.88	0.06	0.02
634	20.70	20.72	20.71	21.72	0.01	0.02
635	18.50	18.49	18.51	18.50	0.02	0.00

TABLE 7—Continued

SRM	Certified Value	% Oxide 1st Run	% Oxide 2nd Run	Average % Oxide	Difference of Duplication	Difference of Average From SRM
636	23.20	23.11	23.15	23.13	0.04	0.07
637	23.10	23.18	23.16	23.17	0.02	0.07
638	21.40	21.39	21.50	21.44	0.11	0.04
639	21.60	21.51	21.61	21.55	0.10	0.05
		CaO (Corrected for Si, Ca, and K ΔM)				
633	64.50	64.75	64.90	64.82	0.15	0.32
634	62.60	62.74	62.61	62.68	0.13	0.08
635	59.80	59.83	59.79	59.81	0.04	0.01
636	63.50	63.52	63.38	63.45	0.14	0.05
637	66.00	66.21	65.85	66.04	0.36	0.04
638	62.10	62.09	61.92	62.00	0.17	0.10
639	65.80	65.58	65.43	65.50	0.15	0.30
		TiO_2 (No corrections)				
633	0.24	0.24	0.24	0.24	0.00	0.00
634	0.30	0.29	0.29	0.29	0.00	0.01
635	0.32	0.33	0.33	0.33	0.00	0.01
636	0.17	0.18	0.18	0.18	0.00	0.01
637	0.21	0.20	0.20	0.20	0.00	0.01
638	0.25	0.25	0.25	0.25	0.00	0.00
639	0.31	0.30	0.31	0.30	0.01	0.01
		Al_2O_3 (Corrected for Al ΔM)				
633	3.74	3.72	3.70	3.71	0.02	0.03
634	5.20	5.12	5.18	5.15	0.06	0.05
635	6.20	6.23	6.18	6.20	0.05	0.00
636	3.10	3.14	3.06	3.10	0.08	0.00
637	3.30	3.30	3.30	3.30	0.00	0.00
638	4.50	4.47	4.40	4.44	0.07	0.06
639	4.30	4.41	4.45	4.41	0.04	0.11
		Fe_2O_3 (Corrected for Fe ΔM)				
633	4.20	4.19	4.19	4.19	0.00	0.01
634	2.87	2.87	2.85	2.86	0.01	0.01
635	2.65	2.73	2.70	2.72	0.03	0.07
636	1.62	1.68	1.68		0.00	
			1.68		0.06	
637	1.80	1.76	1.75	1.76	0.01	0.04
638	3.58	3.60	3.58	3.59	0.02	0.01
639	2.42	2.35	2.35	2.35	0.00	0.07
		K_2O (No Corrections)				
633	0.165	0.161	0.163	0.162	0.002	0.003
634	0.43	0.428	0.427	0.428	0.001	0.002
635	0.45	0.451	0.451	0.451	0.000	0.001
636	0.57	0.572	0.572	0.572	0.000	0.002
637	0.245	0.251	0.251	0.251	0.000	0.006
638	0.59	0.588	0.587	0.588	0.001	0.002
639	0.05	0.049	0.049	0.049	0.000	0.001
		MgO (Corrected for S and Ca ΔM)				
633	1.04	1.09	1.01	1.05	0.08	0.01
634	3.40	3.23	3.53	3.38	0.30	0.02
635	1.25	1.33	1.17	1.25	0.16	0.00
636	4.00	3.98	3.99	3.98	0.01	0.02
637	0.72	0.79	0.82	0.80	0.03	0.08
638	3.84	3.93	3.78	3.86	0.15	0.02
639	1.29	1.24	1.36	1.30	0.12	0.01

TABLE 7—*Continued*

SRM	Certified Value	% Oxide 1st Run	% Oxide 2nd Run	Average % Oxide	Difference of Duplication	Difference of Average From SRM
		SO$_3$ (Corrected for S ΔM)				
633	2.18	2.14	2.14	2.14	0.00	0.04
634	2.16	2.14	2.08	2.11	0.06	0.05
635	7.00	7.03	6.96	7.00	0.07	0.00
636	2.30	2.38	2.37	2.38	0.01	0.08
637	2.33	2.38	2.36	2.37	0.02	0.04
638	2.30	2.28	2.22	2.25	0.06	0.05
639	2.40	2.44	2.41	2.42	0.03	0.02

Suggested Method for X-Ray Emission Spectrometric Analysis of Portland Cement By the Energy-Dispersive Technique

Introduction

This is a standard pressed powder method using an X-ray tube for excitation in an energy dispersive X-ray spectrometer. Nuclear sources are also used with energy dispersive units. In an energy dispersive unit, wavelengths (energies) characteristic of all elements plus scattered radiation from the X-ray source are detected simultaneously by the same detector. Radiation characteristic of the various elements is then separated electronically and measured. The two previous methods used so-called wavelength dispersive optics, whereby the radiation characteristic of each element is physically separated by Bragg diffraction using separate crystals and detectors for each element.

This would appear to give a strong advantage to the energy dispersive approach in that much closer coupling of source-sample-detector can be achieved, resulting in much higher intensities for a given X-ray input than are available with wavelength dispersive optics. While this effect is real, it is not necessarily an advantage because of available detectors. These systems use solid-state lithium-drifted silicon detectors, most of which have to be cooled to liquid nitrogen temperature to operate with good resolution and low noise. Even so, the energy distribution resulting from the lightest elements of interest, specifically sodium through sulfur in the periodic table, overlap and must be mathematically separated based on assumed energy distributions. Another difficulty is that these detectors, as is the case with all detectors now available, saturate at count rates of 20 000 to 50 000 counts per second. That is, energy pulses are received faster than the detector can recover so as to accurately receive and measure the energy of the next pulse. This problem, again, is normally gotten around with a mathematical correction, based on assumed detector performance. A third difficulty is precision. The precision of a measurement in X-ray for a given element is, for large numbers of counts, equal to $\sqrt{1/n}$ where n is the number of counts collected from that element. Calcium oxide constitutes the vast majority, 60 to 70%, of portland cements and clinkers with the oxides of the other elements in lesser proportions. Calcium also has a high excitation potential in the system normally used and is readily detected. Consequently, some 75 to 85% of the total radiation from the sample seen by the detector will be from calcium, with the remainder being from all other elements. In order to obtain adequate precision for minor and trace elements, one must either cut the intensity of the excitation radiation to stay within the optimum operating conditions of the detector and extend analysis time proportionately or provide another mathematical correction based on assumed conditions. A fourth correction to the data is then required to get final analytical results. These are the typical interelement corrections wherein the measured intensities or concentrations are corrected for each of the other elements and their quantities. Usually an iterative procedure will be used, repeating the calculations until the concentrations of each element "close" on a single value; that is, the change resulting from an iteration is negligible. All of these calculations are provided automatically by the analytical system.

This approach works quite well for a sufficiently narrow range of composition as, for example, a cement from a single source, and is highly applicable to routine quality control situations. Investigators have reported difficulties, however, with analysis of unknown samples. One can expect this situation to improve over the next several years with the rapid advances in solid-state technology. Improved detectors with better energy resolution and faster response eventually should eliminate the need for so many assumptions, each of which has a precision and a bias.

Editor

Appears in *Methods for Analytical Atomic Spectroscopy*, 8th ed., E-2 SM 10-34, American Society for Testing and Materials, 1987, pp. 992–996.

Suggested Method for
X-RAY EMISSION SPECTROMETRIC ANALYSIS OF PORTLAND CEMENT BY THE ENERGY-DISPERSIVE TECHNIQUE[1]

SUBMITTED BY B. D. WHEELER[2]

1. Scope

1.1 This method covers the X-ray spectrometric analysis of portland cement for the following constituents in the ranges indicated:

Constituents	Concentration Range, %
Calcium oxide	59.0 to 66.0
Silicon oxide	18.0 to 24.0
Aluminum oxide	3.0 to 7.0
Sulfur trioxide	2.0 to 7.0
Ferric oxide	1.0 to 6.0
Magnesium oxide	0.5 to 4.0
Sodium oxide	0.1 to 1.0
Potassium oxide	0.04 to 1.0
Titanium dioxide	0.1 to 0.4
Phosphorus pentoxide	0.05 to 0.3
Strontium oxide	0.05 to 0.35
Manganic oxide	0.04 to 0.15

NOTE 1 – This method depends on the determination of the elements, but convention of expressing composition in terms of oxides is followed.

2. Applicable Documents

2.1 *ASTM Standards:*

E 50 Recommended Practices for Apparatus, Reagents, and Safety Precautions for Chemical Analysis of Metals[3]

E 135 Definitions of Terms and Symbols Relating to Emission Spectroscopy[4]

3. Summary of Method

3.1 Ground and pelletized samples and standards are placed in a twelve-position sample holder in an energy-dispersive X-ray spectrometer and exposed to an X-ray beam from a rhodium target X-ray tube. The intensities of the secondary X rays generated in the sample are measured by a solid state silicon detector and analyzed through a multichannel analyzer. The output of the detector is integrated for a fixed time. The integration, in counts per unit of time, is stored in a computer. The count rates of the standards are converted to an analytical curve using interelement corrections and a least-squares fit program. The samples are also processed on the computer with interelement corrections and then fitted to the standard calibration curve using a least-squares linear fit.

4. Definitions

4.1 Refer to Definitions E 135.

5. Apparatus

5.1 *Sample Preparation Equipment:*

5.1.1 *Rotary Swing Mill*, with tungsten carbide vial.[5]

5.1.2 *Hydraulic Press*.[5]

5.2 *Excitation Source:*

5.2.1 *X-ray Generator*, with full-wave rectified power supply capable of producing 5 to 50 kV at 1 to 200 μA current.[6]

5.2.2 *X-ray Tube* – High-purity, rhodium target with a 0.25-mm beryllium window.[6]

5.3 *Energy-Dispersive Spectrometer:*

5.3.1 *Detector*, solid-state silicon, 165 eV FWHM at 10 μs shaping time and 1000 counts/s at 5.9 keV Mn K$_\alpha$ peak (^{55}Fe source with 0.008-mm beryllium window).

---

[1] This suggested method has no official status in the Society, but is published as information only. The method is based on the experience of the submitter.
[2] ORTEC Inc., 100 Midland Road, Oak Ridge, Tenn. 37830.
[3] *Annual Book of ASTM Standards*, Part 12.
[4] *Annual Book of ASTM Standards*, Part 42.
[5] Spex Industries, P. O. Box 798, Metuchen, N.J. 08840, or equivalent.
[6] ORTEC 6100 Tube Excited Fluorescence Analyzer, or equivalent.

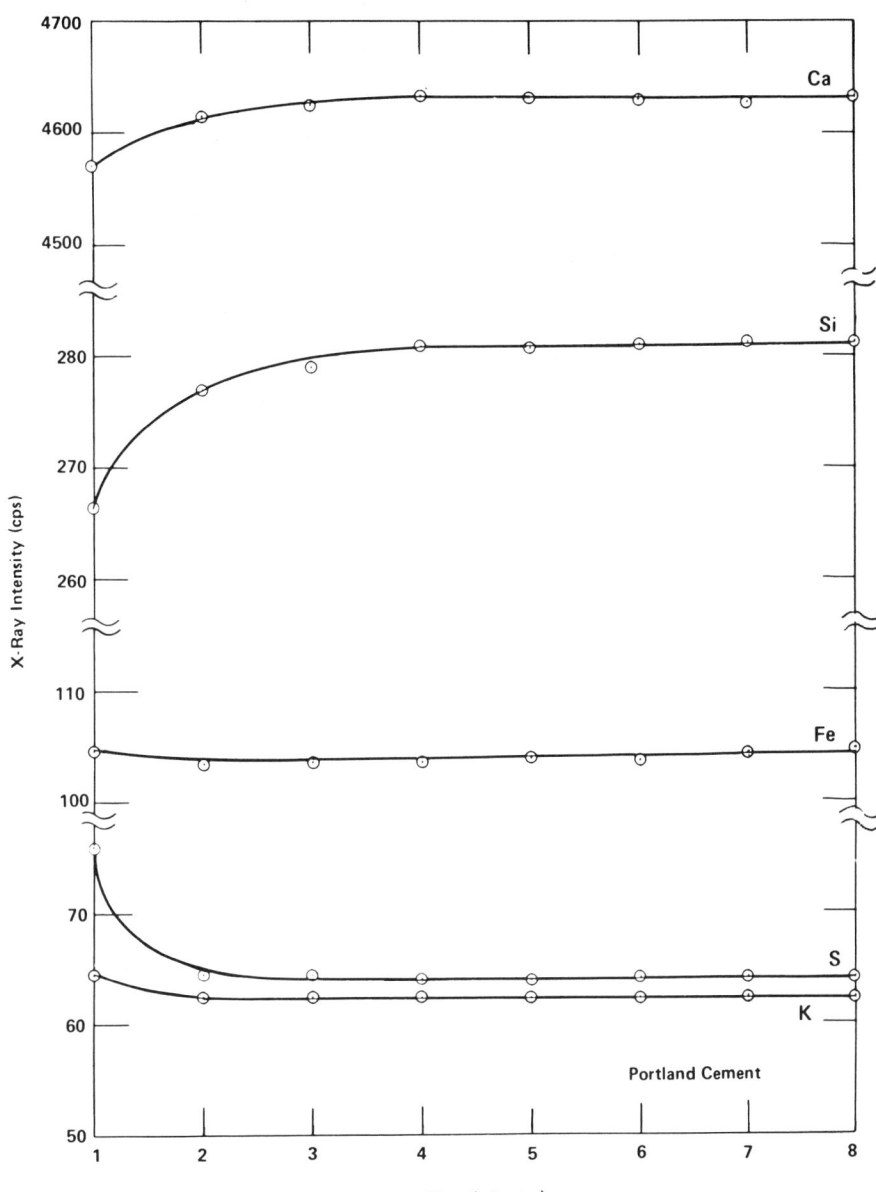

FIG. 1 Grinding Time versus X-ray Intensities.

5.3.2 *Vacuum System*, capable of maintaining the vacuum at a pressure of 200 μm or less of mercury.

5.3.3 *Multichannel Analyzer*, capable of amplifying, counting, and integrating pulses received from the detector.

5.3.4 *Collimator*, 3-mm, for limiting the characteristic X rays to a parallel bundle.

5.4 *Measuring System:*

5.4.1 *Minicomputer*,[7] with 16 K of memory.

5.4.2 *Input/Output Device*.[8]

NOTE 2 — In the absence of a minicomputer, the measurements and calculations can be done manually.

6. Reagents and Materials

6.1 *Purity and Concentration of Reagents* — The purity and concentration of chemical reagents shall conform to the requirements prescribed in Recommended Practices E 50.

6.2 *Grinding Aid*.[9]

6.3 *Ethanol*, anhydrous.

6.4 *Liquid Nitrogen*.

7. Safety Precautions

7.1 Monitoring devices such as film badges or dosimeters should be worn by all personnel operating the equipment. The X-ray spectrometer must have a light to indicate "X rays on" and an interlock system which turns X rays off if the sample chamber is opened while the tube is energized.

8. Standard Reference Materials

8.1 Standard reference materials NBS Nos. 633, 635, 636, 637, 638, 639, and 1016 shall be used for calibration of the X-ray spectrometer.[10]

9. Preparation of Standards and Samples

9.1 *Weighing* — Weigh 5.000 ± 0.005 g of cement and 0.100 ± 0.005 g of grinding aid.

9.2 *Grinding* — At initial calibration, a sample grinding time should be established. Weigh out seven separate samples as in 9.1 from one homogenized cement, place in rotary swing mill, and grind for 1 min; brush sample from mill and pelletize as described in 9.3. Repeat the grinding and pelletizing for the remaining weighed samples for 2, 3, 4, 5, 6, and 7 min. Plot intensity versus grinding time as illustrated in Fig. 1. Observe the time where the intensities stabilize and add 1 min to establish grinding time. Grind all standards and samples for this established grinding time.

NOTE 3 — Establishment of grinding time is necessary since the grinding efficiency between various rotary swing cells is not equal.

9.3 *Pelletizing* — After the sample and grinding aid has been ground for the time established in 9.2, remove the material from the mill with the aid of a brush and transfer to a clean sheet of paper. Pour the ground material into the pellet die and level the powder with a spatula. Place approximately 11 g (two teaspoons) of boric acid on top of the sample in the die. Place the pellet die assembly in a hydraulic press and pelletize the sample at a pressure of 210 MPa (30 000 psi). Hold the pressure for 30 s, release pressure, and remove pelletized sample from the die.

9.4 *Cleaning Mill and Die* — After preparing a sample, clean the mill and die to avoid contamination with the next sample. Brush out the remaining powder residue from the mill and wipe out any remaining traces of powder with a paper towel wetted with ethanol. Clean the die assembly in a similar manner.

9.5 *Preparation of Standards* — Prepare all of the standards listed in 8.1 as described in 9.1 to 9.4. In addition, prepare the well-homogenized sample used in the grinding time study.

10. Preparation of Spectrometer

10.1 *Operating Parameters* — The spectrometer operating parameters are listed in Table 1. Prior to operation, have the X-ray tube in the standby position for 60 min for stability.

NOTE 4 — Follow the X-ray spectrometer manufacturer's instructions for setting up and operating the instrument.

11. Calibration and Standardization

11.1 *Calibration* — Load the prepared standard pellets in the sample chambers 2 through

---

[7] Digital Equipment Corp, PDP 11/05, or equivalent.
[8] Digital Equipment DEC Writer, or equivalent.
[9] A suitable grinding aid is Ivory Snow detergent, manufactured by Procter and Gamble.
[10] Available from the National Bureau of Standards, U. S. Department of Commerce, Washington, D.C. 20234. A new certificate is being prepared for SRMs 633 through 639.

TABLE 1 Analytical Lines and Operating Parameters

Element	keV[A]	Tube Power kV/μA	Filter
Ca	3.69	10/100	none
Si	1.74	10/100	none
Al	1.48	10/100	none
S	2.31	10/100	none
Fe	6.40	10/100	none
Mg	1.25	10/100	none
Na	1.04	10/100	none
K	3.31	10/100	none
Ti	4.51	10/100	none
P	2.01	10/100	none
Sr	14.31	40/200	Mo
Mn	5.90	10/100	none

[A] The K_α lines are used.

TABLE 2 Data on Precision

Oxide	Average Concentration, %	Intralaboratory Coefficient of Variation
Calcium oxide	66.13	0.20
Silicon oxide	23.16	0.78
Aluminum oxide	3.35	2.09
Sulfur oxide	2.31	2.16
Ferric oxide	1.79	0.56
Magnesium oxide	0.73	21.92
Sodium oxide	0.13	...
Potassium oxide	0.25	12.00
Titanium dioxide	0.23	4.35
Phosphorus pentoxide	0.21	14.29
Strontium oxide	0.094	21.28
Manganic oxide	0.060	6.67

9 and the pelletized, homogenous sample in chamber 1. Operate the instrument as specified and establish a calibration curve for the analyte determination.

11.1.1 *Interelement Corrections* — The calibration curve is a linear regression fit of corrected X-ray intensity measurements. The method which has proved reliable is illustrated in the following equation.

$$C = A + BI_c \exp(m_1 I_1 + m_2 I_2, m_3 I_3 + \ldots + m_n I_n) \quad (1)$$

where:
C = percent of element present,
A = the x intercept,
B = the slope of the calibration curve,
I_c = radiation energy intensity of element C,
m_1, m_2, m_3 = positive or negative constants of interfering elements 1, 2, and 3 derived from the calibration standards, and
I_1, I_2, I_3 = radiation energy intensities of interfering elements 1, 2, and 3.

These calculations are performed automatically by the computer interfaced to the spectrometer.

NOTE 5 — The purpose of using an homogenized pellet with the standards is to establish intensities during initial calibration which will be used to correct for any electron drift during succeeding analyses of samples (see 12.1).

12. Analysis of Samples

12.1 *Analysis Using Calibration* — Place the previously analyzed homogenized pellet in chamber 1 and the samples in chambers 2 through 12.

12.2 *Correction for Electronic Drift* — The radiation from the homogenized pellet is counted in an identical manner as were the standards. The electronic drift is compensated by ratioing the intensities of the homogenized sample at the data of calibration to the intensities of the same sample as follows:

$$\frac{I_h \text{ at date of calibration}}{I_h \text{ at time of analysis}} \times I_i \text{ of sample} = I_c \text{ or corrected } I_i \quad (2)$$

where:
I_h = radiation energy intensity of element in homogenized pellet, and
I_i = radiation energy intensity of element in sample at time of analysis.

12.2.1 Make the electronic drift correction for each element.

13. Calculations

13.1 By means of the computer interfaced to the spectrometer, calculate the concentration of each constituent substituting the X-ray intensity which has been corrected for possible electronic drift for that element into Eq (*1*).

14. Precision and Accuracy

14.1 *Precision* — The precision of the X-ray measurements was determined by preparing several samples on different days and analyzing each sample several times on different days. The data shown in Table 2 represent single-operator, multi-day precision.

14.2 *Accuracy* — Three samples were analyzed and reported. The true values were subsequently provided and are reported in Table 3.

TABLE 3 Data on Accuracy[a]

Oxide	Sample No. 1		Sample No. 2		Sample No. 3	
	Percent Present	Percent Found	Percent Present	Percent Found	Percent Present	Percent Found
Calcium oxide	66.97	67.13	62.24	62.34	64.30	64.42
Silicon dioxide	20.51	20.41	22.25	22.33	21.46	21.37
Aluminum oxide	5.51	5.56	5.15	5.20	4.24	4.31
Sulfur trioxide	1.86	1.95	1.76	1.88	1.42	1.50
Ferric oxide	2.43	2.43	3.16	3.16	5.40	5.46
Magnesium oxide	0.84	1.04	3.05	2.95	0.90	1.02
Sodium oxide	0.06	0.08	0.26	0.23	0.60	0.55
Potassium oxide	0.25	0.27	0.70	0.70	0.14	0.14
Titanium dioxide	0.26	0.24	0.26	0.23	0.28	0.25
Phosphorus pentoxide	0.40	0.36	0.09	0.08	0.26	0.23
Strontium oxide	0.08	0.07	0.18	0.17	0.14	0.15
Manganic oxide	0.04	0.04	0.09	0.11	0.05	0.05

[a] Samples were provided by the Portland Cement Association. The analyses were provided after the determinations by energy-dispersive X ray had been completed.

Section III
Spectrophotometric/EDTA Methods

Only one set of procedures is included here. This is the scheme of analysis for hydraulic cements and clinker devised by Z. T. Jugovic while he was with Universal Atlas Cement (formerly, Division of the United States Steel Company). These methods are an updated version of those published in *Analytical Techniques as Applied to Hydraulic Cements and Concrete, ASTM STP 395*, 1966, pp. 65–93. For ease of reference, these have long been denoted as the Jugovic Procedures. They were developed and are suitable for portland, blended, and calcium aluminate cements and clinkers. The procedures can also be applied, with appropriate modifications, to raw meal and raw materials.

Editor

Z. T. Jugovic[1]

Spectrophotometric and EDTA Methods for Rapid Analysis of Hydraulic Cement

Introduction

These methods have been used successfully on a routine basis by at least a dozen laboratories. Their primary advantage over the other rapid methods that provide schemes for "complete" analyses is much lower equipment cost. Methods are included for analysis of silica (SiO_2), alumina (Al_2O_3), ferric oxide (Fe_2O_3), calcium oxide (CaO), magnesium oxide (MgO), titanium oxide (TiO_2), phosphorus pentoxide (P_2O_5), manganese sesquioxide (Mn_2O_3), and sulfur trioxide (SO_3). Methods are not provided for sodium oxide (Na_2O) and potassium oxide (K_2O); these normally would be run by flame photometry.

Analysis time is about the same to a little longer than that required for atomic absorption, but is considerably longer than for X-ray. Still, an experienced analyst can analyze several samples per normal work day and still have time for other duties.

Editor

1. Scope

1.1 These methods cover the spectrophotometric and EDTA procedures for rapid chemical analysis of hydraulic cements.

2. Apparatus and Materials

2.1 *Spectrophotometer*—Beckman Model B spectrophotometer with matched 1-cm glass cells and photomultiplier attachment or equivalent (Note 1).

NOTE 1—Present methods were developed using Beckman Model B spectrophotometer, and therefore settings and the operations described pertain specifically to that instrument. Other comparable spectrophotometers may be used.

2.1.1 All spectrophotometric readings are made on the absorbance scale. The instrument settings should permit absorbance readings to ±0.001.

2.1.2 The instrument setting and absorbance ranges for individual measurement are listed in Table 1 as a general guideline of the values to be expected. The actual values will vary from one instrument to another.

2.1.3 The wavelengths must be set very carefully, particularly for those determinations where measurements are not made at the wavelength of maximum absorbance (for example, 410 nm for SiO_2).

---

[1] 1324 Beatrice Lane, Munster, IN 46321.

TABLE 1—*Settings, absorbance ranges, and concentrations used with Beckman Model B spectrophotometer.*

Constituent	Phototube	Wavelength, nm	Sensitivity Switch Position	Absorbance Range	Concentration	
					mg/mL[a]	Corresponding Range in Original Sample, %
SiO_2	Photomultiplier Voltage A	410	2	0.050 to 1.170	0 to 4.6/100	0 to 24.0
Al_2O_3 (R_2O_3)	"	365	2	0.500 to 1.140	0 to 0.448/100	0 to 7.0
Fe_2O_3	"	600	2	0.005 to 0.300	0 to 0.488/100	0 to 7.0
SO_3	"	450	2	0.125 to 0.910	0.87 to 3.75/50	0.7 to 3
Mn_2O_3	"	525	2	0.030 to 0.620	0.1 to 2/100	0.01 to 0.20
TiO_2	"	410	2	0.030 to 1.070	0 to 0.32/50	0 to 1.0
P_2O_5	red	725	2	0.002 to 1.200	0 to 0.32/50	0 to 0.5

[a] Final concentration per volume on which the absorbance measurement is made.

2.2 Voltage Regulating Transformer—Beckman 12320, 60 cycles, AC, or any other comparable transformer.

2.3 Crucible—A pure gold crucible with reinforced bottom, 70-mL capacity and a gold lid.

3. Reagents

3.1 Distilled water or deionized water of comparable purity must be used throughout the analysis. Reagent-grade chemicals should be used except when no reagent grade is available (for example, for ferron and phthalein purple):

- acetic acid, glacial (CH_3COOH)
- aluminum chloride, crystal ($AlCl_3 \cdot 6H_2O$)
- aluminum potassium sulfate, crystal ($Al_2(SO_4)_3 \cdot K_2SO_4 \cdot 24H_2O$)
- ammonium chloride (NH_4Cl)
- ammonium hydroxide, sp gr 0.90 (NH_4OH)
- ammonium molybdate, crystal (($NH_4)_6Mo_7O_{24} \cdot 4H_2O$)
- ascorbic acid, powder ($C_6H_8O_6$)
- barium chloride, crystal ($BaCl_2 \cdot 2H_2O$)
- calcium carbonate ($CaCO_3$), primary standard
- cement, National Bureau of Standards (NBS) Nos. 1011 and 1013
- Disodiumethlenediaminetetraacetatedihydrate ($Na_2C_{10}H_{14}N_2O_8 \cdot 2H_2O$), EDTA
- ferric chloride ($FeCl_2 \cdot 6H_2O$)
- ferron (8-hydroxy-7-iodo-5-quinolinesulfonic acid)
- ferrous ammonium sulfate, crystal ($FeSO_4(NH_4)_2SO_4 \cdot 6H_2O$)
- hydrochloric acid (HCl), sp gr 1.19
- hydroxylamine hydrochloride ($NH_2OH \cdot HCl$)
- hydroxy naphthol blue indicator, Mallinckrodt No. 5630
- naphthol green B indicator, $[C_{10}H_5(NO)(SO_3Na)O]_3Fe$, Allied Chemical, National Aniline Division Catalogue No. 670
- nitric acid (HNO_3), sp gr 1.42
- methyl red indicator, (1) $(CH_3)_2NC_6H_4N:NC_6H_4COOH$
- molybdic acid (assay as MoO_3) minimum 85%
- phthalein purple indicator, Fisher Scientific Co. No. P-379 or equivalent
- potassium chloride (KCl)
- potassium hydroxide (KOH)
- potassium phosphate, monobasic crystal (KH_2PO_4)
- potassium pyrosulfate ($K_2S_2O_7$)
- sodium acetate ($NaC_2H_3O_2 \cdot 3H_2O$)
- sodium hydroxide (NaOH), pellets (assay min 98%)
- sodium sulfate (Na_2SO_4), anhydrous powder
- sulfuric acid (H_2SO_4), sp gr 1.84
- tiron, indicator (($HO)_2C_6H_2(SO_3Na)_2$)
- titanium dioxide (TiO_2), NBS No. 154a or equivalent
- triethanolamine (2,2′,2″-nitrilotriethanol, $(CH_2OHCH_2)_3N$)

Specific reagent preparations are given under respective procedures.

4. Preparation of Sample Solution 1 for the Determination of SiO_2, Al_2O, Fe_2O_3, CaO, MgO, TiO_2, and P_2O_5

4.1 Reagents

4.1.1 Hydrochloric acid 5.50 ± 0.05N

4.1.2 Sodium hydroxide (pellets, assay 98.0% minimum); store in tightly closed, plastic bottle.

4.2 Procedure

4.2.1 Weight 5.1 ± 0.1 g (or one pellet) sodium hydroxide into a 70-mL gold crucible (Note 2), then add 0.3200 ± 0.0001 g cement (Note 3). Cover and place the crucible in a muffle furnace on a raised clay or fused silica triangle so that the bottom of the crucible does not touch the furnace refractory plate. Heat at 800°C for 5 min. Remove from the furnace and gently swirl until the melt beings to solidify. Cool the outside of the crucible by dipping several times the lower 1/3 of the crucible into cold distilled water. (Caution: Do not allow any water to come in contact with the contents of the crucible to avoid violent reaction of the molten sodium hydroxide [NaOH] and water). Place the still warm crucible and the lid in a 600-mL plastic beaker marked at 150 mL. Pour 100 mL of vigorously boiling water into the beaker, and immediately cover with a watch glass. The fused mass disperses rapidly in boiling water. When the vigorous reaction is over, after about 1 min, wash the watch glass, with the gold tipped tongs remove and carefully rinse the crucible, and the cover with hot water. Keep the volume in the plastic beaker just below the 150-mL mark.

NOTE 2—Cover crucible containing NaOH while weighing cement sample to prevent moisture absorption.

NOTE 3—If the sample contains more than 24% SiO_2 use a smaller sample weight that will contain an equivalent amount of SiO_2 as a 0.3200-g sample of another cement having 20 to 22% of SiO_2. In such a case, also make necessary adjustments in the calculation of the results. For example, if a Type IS cement has about 28% SiO_2, a 0.2400-g sample should be used for SiO_2 analysis and the results multiplied by 0.3200/0.2400 = 1.333.

4.2.2 Measure 50.0 mL of 5.50 N hydrochloric acid (HCl) into a 250-mL volumetric flask. To keep the alkaline-water leach in suspension, swirl the contents of the plastic beaker, then pour through a long-stemmed plastic funnel into the acid, swirling the flask to mix rapidly. With a fine stream of hot water from a wash bottle, rinse the plastic beaker twice, keeping the total volume of the rinse water below 20 mL. Add 5-mL 5.50 N HCl to the plastic beaker, swirl, and then pour into the gold crucible. Cover and heat the crucible while rinsing the plastic beaker with water into the main solution. Warm the acid solution in the flask for 1 min. Add the acid from the crucible to the main solution, rinsing the crucible with a small amount of hot water. Allow the flask to cool on the bench for a few seconds, then cool to room temperature in running water (2 min). Place the flask in the constant temperature bath (24 ± 1°C) for 5 to 10 min, dilute to mark, and mix well. This is Solution 1. Determine silica within 2 h.

4.3 Blank—Prepare a fusion blank solution in the same manner. A fresh fusion blank solution should be prepared each time new batches of reagents are used. Otherwise aliquots of the same blank solution can be used for several days as the blank for Al_2O_3, Fe_2O_3, TiO_2, and P_2O_5.

Silicon Dioxide

5. Reagent

5.1 Molybdic Acid Reagent (10% in 2.5 N HNO_3). Mix 240-mL (HNO_3) (sp gr 1.42) with 260 mL of water, and set aside to cool in running cold water.

5.1.1 Dissolve 117.6 g of molybdic acid (85%) in a solution of 420-mL water plus 52.7-g NaOH (98.0%). Filter through Whatman 541 paper, or equivalent, and cool in running cold water. Pour slowly into cold nitric acid solution while stirring. (Any cloudy precipitation forming must be completely dissolved before adding more molybdic acid to the nitric acid. Keep nitric acid solution in cold water during mixing.) Cool, filter under suction using Pyrex filter holder[2] and glass fiber filter paper, transfer to a polyethylene bottle with cap and store in a refrigerator at approximately 5°C to prevent molybdic-acid precipitation.

NOTE 4—Leave the molybdic-acid reagent in refrigerator 6 to 8 h, preferably overnight, before calibration.

NOTE 5—At 5°C this supersaturated solution of molybdic acid usually remains stable in excess of six weeks.

NOTE 6—Do not shake or pour the solution out of the bottle as shaking and incrustation around the neck of the container will tend to speed up precipitation of molybdic acid and thereby weaken the reagent. When measuring the required amount of this reagent for subsequent SiO_2 determination, use a clean, dry pipet to withdraw the reagent directly from the plastic bottle. As soon as the required amount of the reagent is measured, place the remaining reagent again under refrigeration.

6. Calibration

6.1 Fuse triplicate 0.3200 ± 0.0001 g samples of NBS Cement 1011 or equivalent, and prepare to 250-mL volume as described in Section 4.2. Also prepare one blank solution using the same procedure but omitting the sample. Using a buret, transfer 2 aliquots from each sample solution into separate 100-mL volumetric flasks as follows: one 15-mL sample aliquot (equivalent to 21.03% SiO_2 in NBS No. 1011 cement) and one 12.5-mL sample aliquot (equivalent to 17.53% in SiO_2 NBS No. 1011 cement) plus 2.5-mL fusion blank. Add 70-mL water and continue as described in Sections 7.2 through 8.1

6.2 Also prepare fused samples of NBS Cement 1013 or equivalent in triplicate. Determine SiO_2 on a 15-mL aliquot in the usual way (SiO_2 = 24.17% in NBS No. 1013).

6.3 If the difference between extreme net absorbance reading values within a triplicate set is greater than 0.010, re-analyze that sample until the three values agree within 0.010 absorbance, then average the three values. Plot mean (net) absorbance for each triplicate set of readings against percent SiO_2 on the scale of 1 cm = 0.1% SiO_2, and 1 cm = an absorbance increment of 0.010. Draw a smooth curve through the points. (Suitable graph paper is 50 cm wide.)

---

[2] Catalogue No. XX 1004700 manufactured by Millipore Filter Corp., Bedford, MA.

7. Procedure

7.1 Measure with a buret 15 mL of Solution 1 into a 100-mL volumetric flask, and add 70-mL water.

7.2 Add 10-mL molybdic acid with a pipet, swirling the flask during addition. Make up to near-volume, mix, and immerse in a constant-temperature bath at 24 ± 1°C.

7.3 After 20 ± 1 min, make up to volume, mix, and measure absorbance to the nearest 0.002 at exactly 410 nm.

CAUTION—The wavelength setting must be reproduced very carefully, because the absorbance changes rapidly as the wavelength is changed only slightly. A change of 1 nm alters the absorbance by about 0.04, equivalent to 0.9 percentage points of SiO_2.

7.4 After an additional 5 min in the constant-temperature bath, measure absorbance again to ascertain that the maximum color intensity is reached. If necessary, repeat step 7.4.

NOTE 8—It is advisable to measure when the absorbance is near its maximum. The time needed to reach maximum absorbance depends on the amount of silica in the sample, acidity, and temperature. Usually, maximum absorbance is reached after 20 min for the samples containing 20 to 22% SiO_2. After reaching maximum absorbance, the color slowly fades (at about 0.01% SiO_2/min).

8. Calculation

8.1 Substract cell correction from the maximum absorbance reading to obtain net absorbance.

8.2 Read percent SiO_2 from standard calibration graph.

NOTE 8—In order to ascertain that the molybdic-acid reagent has not changed since the time of calibration, it is recommended to run NBS Cement 1011 or equivalent for SiO_2 in the same manner. If the SiO_2 result is within 0.20 percentage points of the certified value, no correction is needed for the SiO_2 of the unknown sample. If the SiO_2 value of the NBS cement sample deviates by more than 0.20 percentage points, run another SiO_2 determination on the same, but freshly fused NBS standard cement. If the SiO_2 still deviates by more than 0.20 percentage points from the certified value, correct the percent SiO_2 of the unknown sample by the difference between the average of the obtained values of the NBS cement and the certified SiO_2 value. In such a case, the additional unknown samples should not be analyzed with the same molybdic-acid reagent. It was found from practical experience over several years that the molybdic-acid reagent is stable in excess of six weeks. When determining SiO_2 with a molybdic-acid reagent older than six weeks, a one-a-day companion NBS cement should be analyzed as a precautionary measure.

NOTE 9—When a new molybdic-acid reagent is made, check the midpoint of the calibration curve with 15-mL aliquots from three fresh fusions of the NBS Cement 1011 or equivalent. The difference in the mean absorbance at the midpoint of the calibration curve obtained with the fresh molybdic-acid reagent, and that obtained at the time of initial calibration is then applied as a correction (positive or negative) for the absorbance of the unknown sample before reading the percent SiO_2 from the calibration curve.

Aluminum Oxide and Ferric Oxide

9. Reagents

9.1 Ferron (8-hydroxy-7-iodo-5-quinolinesulfonic acid)—(0.2%[3] in 2.5 N acetate buffer). Dissolve 370-g hydrated sodium acetate ($CH_3COONa \cdot 3H_2O$) in 800-mL warm water. Filter through a Whatman 541 or equivalent filter paper, and add 470-mL standardized 5.0 N acetic acid.

9.1.1 Dissolve 4.00 g of ferron in 100-mL water containing 1-g NaOH. (The dissolution of ferron is greatly facilitated if only a small amount of sodium hydroxide solution is added first to make a paste while mixing with a glass rod.) Add ferron solution to the sodium acetate solution to make 2000 mL, mix, and transfer to a dark-colored bottle. Let it stand for a few hours (preferably overnight), and then filter using a Pyrex filter holder and glass fiber filter paper. Determine the pH with a pH meter, and adjust to 4.7 ± 0.05. (Usually a small addition of HCl is needed to obtain pH 4.7.)

9.2 Stock Fe_2O_3 Solution—Weigh 0.1568-g ferrous ammonium sulfate ($FeSO_4 \cdot (NH_4)_2 \cdot SO_4 \cdot 6H_2O$); add 5-mL HCl and two drops of HNO_3; cover and heat until the solution changes from green to deep orange. Transfer quantitatively into a 250-mL volumetric flask. Add 13 mL of 5.5 N HCl, cool to room temperature, and dilute to mark. This stock solution contains 128-ppm Fe_2O_3.

9.3 Stock Al_2O_3 Solution—Dissolve in water 0.2975-g aluminum potassium sulfate ($Al_2(SO_4)_3 \cdot K_2SO_4 \cdot 24H_2O$); add 21-mL 5.5 N HCl and dilute to 250 mL in a volumetric flask. This stock solution contains 128-ppm Al_2O_3.

9.4 Hydrochloric Acid—0.05 N.

10. Calibration

10.1 Transfer 50.0-mL Al_2O_3 stock solution into a 500-mL volumetric flask, and 50.0-mL Fe_2O_3 stock solution into another 500-mL volumetric flask. Dilute both solutions to mark. A 10-mL aliquot of either, taken for the usual colorimetric R_2O_3 determination, is equivalent to 2% of the respective oxide in a cement.

10.2 Make up four solutions containing 15, 30, 35, and 40 mL of dilute Al_2O_3 solution; and three solutions containing 10, 25, and 40 mL of dilute Fe_2O_3 solution. Dilute each to 70 mL with 0.05 N HCl, pipet 20-mL ferron, and dilute to 100 mL with 0.05 N HCl. These solutions are equivalent to 3, 6, 7, and 8% Al_2O_3, and 2, 5, and 8% Fe_2O_3, respectively, Also prepare the reagent blank (70-mL 0.05 N HCl + 20 ferron), dilute to 100 mL with 0.05 N HCl, mix thoroughly, and let stand 20 to 30 min.

10.3 Read absorbance of all eight solutions at 365 nm. Also read the absorbances at 600 nm of the three Fe_2O_3 solutions and the reagent blank. Correct for blank at both wavelengths. Repeat the calibration procedure two more times.

10.3.1 If the difference between extreme net absorbance reading values within a triplicate set is greater than 0.010 for the readings at 365 nm and greater than 0.005 for the readings at 600 nm, reanalyze that sample until the three values agree within the specified limits, and then calculate the mean absorbances.

---

[3] Ferron reagent from different manufacturers vary in strength, and the concentration of this solution may require adjustment in order to obtain the recommended absorbance of 500 for the blank.

10.3.2   Plot mean (net) absorbances of Fe_2O_3 solutions at 365 and 600 nm against percent Fe_2O_3 on a scale of 1 cm = 0.10% Fe_2O_3 and 1 cm = 0.010 absorbance. Draw a smooth curve through the points at 600 nm (Curve A) and another through the points at 365 nm (Curve B).

10.3.3   Plot mean (net absorbances of Al_2O_3 solutions at 365 nm against percent Al_2O_3 on a scale of 1 cm = 0.20% Al_2O_3 and 1 cm = 0.020 increment in absorbance. Draw a smooth curve through the points at 365 nm (Curve C).

10.4   When a fresh reagent solution is made, recalibrate.

11. Procedure

11.1   Transfer with a buret a 5-mL aliquot of Sample Solution 1 to a 100-mL volumetric flask and add 70-mL water with a graduated cylinder.

NOTE 10—If the Al_2O_3 in the sample exceeds 7%, or if the Al_2O_3 + Fe_2O_3 is greater than 9%, use a smaller aliquot and make up the difference in volume with the fusion blank (for example, 4 mL of Sample Solution 1 plus 1-mL fusion blank). In such case, also make necessary adjustments in the calculation of the results.

11.2   Prepare a blank in the same manner except with the fusion blank instead of the sample solution.

11.3   Add with a pipe, 20-mL ferron reagent to each flask, dilute to volume with distilled water, mix thoroughly, and let stand about 20 to 30 min.

11.4   Preset the spectrophotometer at 365 nm, and then measure absorbance at 365 nm to the nearest 0.002. Leave the sample and the blank in the spectrophotometer while adjusting the instrument for measurements at 600 nm, and then measure absorbance of the same sample at 600 nm. Correct the readings for blank (and the cell correction) to obtain net absorbances.

12. Calculation

12.1 Fe_2O_3—Read percent Fe_2O_3 from a curve drawn from the net absorbance at 600 nm (Curve A).

12.2 Al_2O_3—Aluminum oxide is determined by either of the following methods:

12.2.1   Determine the percent of Fe_2O_3 from Curve A (the calibration curve for iron at 600 nm). Then read equivalent absorbance at 365 nm from Curve B (the calibration curve for iron at 365 nm). Subtract the value read on Curve B from net absorbance value of the sample at 365 nm, and then read from Curve C (alumina calibration curve) the percent Al_2O_3. Correct the percent of Al_2O_3 for titanium dioxide (TiO_2) by subtracting 0.8 times TiO_2 to obtain true Al_2O_3.

12.2.2

$$Al_2O_3 = \{[A_{365} - (A_{600} \times F_1)] \times F_2\} - (0.8 \times TiO_2\%)$$

where

A_{365} = absorbance reading of the sample at 365 nm (net),
A_{600} = absorbance reading of the sample at 600 nm (net),
F_1 = iron conversion factor to obtain equivalent absorbance reading at 365 nm,

F_2 = factor for converting absorbance caused by alumina-ferron complex to percent Al_2O_3, and

0.8 = TiO_2/Al_2O_3 molecular ratio.

The factors F_1 and F_2 can be calculated from the respective calibration curves.

Calcium Oxide and Magnesium Oxide

13. Stirring Apparatus

13.1 Thermolyne "Stir-Light" magnetic stirrer or equivalent.

14. Reagents

14.1 Standard calcium chloride solution, 0.01 M—Accurately weigh 2.0000 g of dried primary-standard-grade $CaCO_3$ in a 500 mL beaker and add 200 mL of water. While stirring, slowly add 200 mL of HCl (1:1). Boil to expel carbon dioxide, cool to room temperature, and dilute to 2 L. One millilitre of this solution is equivalent to 0.0005608-g CaO.

14.2 EDTA Solution, 0.01 M—Dissolve in water 3.73 g of the disodium salt of EDTA (MW 372.254) per each litre of final solution in water, dilute to volume, and store in a polyethylene bottle. Standardize by the procedure for determination of CaO (Section 15.1) and CaO plus MgO (Section 15.2), using 50 mL of standard calcium chloride solution in place of Sample Solution 1 (standardize daily).

14.3 Buffer pH 12.5 (KOH Solution)—Dissolve 200 g of potassium hydroxide per litre of water, and store it in a closed polyethylene bottle.

14.4 Buffer pH 10 (NH_4Cl–NH_4OH Solution)—Dissolve 68 g of NH_4Cl in about 200 mL of water, add 570 mL of NH_4OH, and dilute to 1 L with water.

14.5 Hydroxy Naphthol Blue Indicator—This is an indicator mixture with KCl and can be used as purchased. It is recommended that the indicator mixture be ground to a powder to avoid segregation.

14.6 Triethanolamine (1:4).

14.7 Hydroxylamine Hydrochloride Solution (10% by weight).

14.8 Phthalein Purple Indicator—Mix intimately by grinding in a porcelain mortar 0.10-g phthalein purple, 0.075-g naphthol green B, 0.005-g methyl red, and 10-g KCl.

15. Procedure

15.1 CaO—Pipet 50-mL aliquot of Solution 1 (equivalent to 0.064 sample) into a 400-mL beaker; place magnetic stirring bar into the beaker and dilute with water to about 150 mL. Place the beaker in the "stir-light" magnetic stirrer and while stirring add in the following order: 5-mL hydroxylamine hydrochloride, 10-mL triethanolamine, and 0.01 M EDTA solution equivalent to about 95% of the calcium present. Raise the pH to 12.5 by adding 20 mL of potassium hydroxide (buffer pH 12.5) and stir for 2 min. Then add about 0.2- to 0.3-g hydroxy naphthol blue indicator, and titrate with 0.01 M EDTA solution until color changes from wine-red to blue. Add 1-mL excess EDTA. Add a measured amount of standard calcium solution to change the color to wine-red. Complete the titration with dropwise addition of EDTA until the first pure blue color appears (Note 11). Correct the volume of EDTA used in titration for the amount of standard calcium chloride added to obtain the volume V_1 needed for titration of calcium in the sample.

NOTE 11—A slight excess of EDTA produces a blue-green coloration. The end point may be checked by adding another measured amount of standard calcium solution and titrating with EDTA. Observe the color change against the diffused fluorescent light of the magnetic "stir-light" stirrer.

15.2 MgO (CaO + MgO)—Pipet a second 50-mL aliquot of Solution 1 into a 400-mL beaker and dilute with water to about 100 mL. While stirring, add in the following order: 5-mL hydroxylamine hydrochloride, 10-mL triethanolamine, and 40-mL buffer pH 10. Add a volume of 0.01 M EDTA equal to the amount of EDTA used for calcium titration, then add about 50-mg (with a measuring scoop delivering 50 ± 5 mg) phthalein purple indicator. Continue the titration until the color changes from purple to bright green. (The color changes from purple to colorless or pale gray, to bright green.) Observe the color change against the diffused fluorescent light of the magnetic "stir-light" stirrer.

16. Calculation

Let

V_1 = volume, mL, of EDTA required for calcium titration in Section 15.1,
V_2 = volume, mL, of EDTA required for calcium plus magnesium titration in Section 15.2,
F_1 = grams CaO/mL EDTA (1-mL EDTA 0.01 M = 0.0005680-g CaO),
F_2 = grams MgO/mL EDTA (1-mL EDTA 0.01 M = 0.0004032 g MgO),
W = weight of sample, and
A = volume of aliquot, mL.

$$\%CaO = \frac{V_1 \times F_1 \times \frac{250}{A}}{W} \times 100$$

$$\%MgO = \left[\frac{(V_2 - V_1) \times F_2 \times \frac{250}{A}}{W} \times 100\right] - (\%MnO \times 0.6)$$

Thus, if $W = 0.32$ g, $F_1 = 0.0005608$, $F_2 = 0.0004032$, $A = 50$ mL, then

$$\%CaO = V_1 \times 0.8763$$

$$\%MgO = [(V_2 - V_1) \times 0.6300] - (\%MnO \times 0.6)$$

Sulfur Trioxide

17. Apparatus

17.1 Stirring Apparatus—A stirrer (a Sargent S-76485 synchronous rotator or equivalent) capable of maintaining a constant speed of 600 rpm.

17.2 Stirring Rod—A stirring rod made of solid glass, 0.5 to 0.6 cm in diameter and about 15 cm in length, bent at the point about 3 cm from the lower end so that the tip describes a circle 2.5 to 3.0 cm in diameter.

17.3 **Measuring Scoop**—A small scoop made from a piece of metal or glass tubing attached to a suitable handle, and of such size as to contain 0.2 g of the barium chloride crystals when filled.

18. Reagents

18.1 **Barium Chloride Crystals**—Use sieved crystals of $BaCl_2 \cdot 2H_2O$ that pass a No. 40 (420-μm) sieve, and are retained on a No. 80 (177-μm) sieve. (Prepare a supply of crystals sufficient for a large number of determinations because a new calibration curve must be prepared for each new batch of barium chloride.)

18.2 **Hydrochloric Acid**, 1.7 ± 0.05 N.

18.3 **Solution B** (0.2500-g SO_3/L)—Dissolve 0.4436-g anhydrous Na_2SO_4 (or 0.5377 g $CaSO_4 \cdot 2H_2O$) in water and dilute to 1000 in a volumetric flask.

19. Calibration

19.1 **Solution A**—To a 2.5-g sample of known-low-SO_3 clinker in a 250-mL beaker, add 50-mL water and 30-mL 5.5 N HCl. Heat to dissolve, and filter through a Munktells lF filter paper into a 200-mL volumetric flask. Rinse beaker, and wash residue five times with hot water. Cool in a water bath to room temperature, and dilute to volume.

19.2 Prepare the following five solutions by measuring specified volumes from three burets and mixing in 100-mL beakers:

1. 10.0-mL Solution A + 2.5-mL Solution B + 37.5-mL water
2. 10.0-mL Solution A + 5.0-mL Solution B + 35.0 mL water
3. 10.0-mL Solution A + 10.0-mL Solution B + 30.0-mL water
4. 10.0-mL Solution A + 10.0-mL Solution B + 25.0-mL water
5. 10.0-mL Solution A + 20.0-mL Solution B + 20.0-mL water

19.2.1 Using the barium chloride to precipitate the sulfate as described in Section 20.2, determine the absorbance at 450 nm. These mixtures represent clinker SO_3 plus additions of 0.5, 1.0, 2.0, 3.0, and 4% SO_3, respectively. Make triplicate determinations with all mixtures and plot the average net absorbance readings against percent of SO_3 on a scale of 1 cm = 0.10 percent SO_3 and 1 cm = 0.020 absorbance.

19.3 Prepare a fresh calibration curve for each new batch of $BaCl_2 \cdot 2H_2O$.

20. Procedure

20.1 To 0.5000 g of cement in a 100-mL beaker, add 5 mL water and 20 mL of 1.7 N HCl. Warm to dissolve, breaking up any lumps with a glass rod if necessary. Transfer to a 200-mL volumetric flask, rinsing the beaker several times with water. Cool to room temperature, dilute to volume, and mix.

20.2 Filter through a dry, 12.5-cm Munktells No. 1F paper, discarding first 20 mL of the filtrate. Pipet 50 mL of the filtrate into a 100-mL beaker (low form). Add 0.2 ± 0.05 g $BaCl_2 \cdot 2H_2O$ (measured with a scoop delivering 0.2 ± 0.05 g $BaCl_2 \cdot 2H_2O$). Wait 5 s, and then stir at 600 rpm for 60 s. Turn stirrer off, transfer the solution into the cell, and measure the absorbance of the cloud at 450 nm and 120 s after beginning to stir.

21. Calculation

21.1 Read the percentage of SO_3 from the calibration curve.

NOTE 12—If the reading falls above the limits of the calibration curve, the determination may be repeated with a smaller amount of the sample and correction made for the difference in weight.

Titanium Dioxide

22. Reagents

22.1 Tiron (disodium-1, 2-dihydroxybenzene-3, 5 disulfonate), 4% water solution. Prepare fresh daily.

22.2 EDTA (0.2 M)—Dissolve 37.5-g EDTA (disodium salt) in 350-mL warm water, and filter. Add 0.25-g $FeCl_3 \cdot 6H_2O$ and dilute to 500 mL.

22.3 Hydrochloric Acid (1.7 N)—Same as for determination of SO_3.

22.4 Buffer (pH 4.7) 68 g $NaC_2H_3O_2 \cdot 3H_2O$, plus 380-mL water, plus 100-mL 5.0 N CH_3COOH.

22.5 NH_4OH (1 to 1).

22.6 Stock TiO_2 Solution—Fuse in a platinum crucible 0.0402 g of NBS Standard 154a (TiO_2), or the appropriate amount of reagent TiO_2 of known composition, with about 2 to 3 g of $K_2S_2O_7$. Allow to cool, and place the crucible in a beaker containing 125-mL H_2SO_4 (1:1). Heat and stir until the melt is completely dissolved. Cool, and dilute the solution to 250 mL in a volumetric flask.

22.7 Dilute TiO_2 Standard Solution—Pipet 50 mL of stock TiO_2 solution into a 500-mL volumetric flask, and dilute to volume. One mL of this solution is equal to 0.016-mg TiO_2, which is equivalent to 0.05% TiO_2 when used as outlined in 23.

23. Calibration

23.1 With a buret, transfer into 50-mL volumetric flasks 0, 5, 10, 15 and 20 mL of dilute TiO_2 standard solution (Section 22.7). Dilute each to 25 mL with water, add 5 mL Tiron, and so forth, continue as outlined in Section 24.2. Based on the original 0.32-g sample, these standards represent 0.0, 0.25, 0.50, 0.75, and 1.00% TiO_2.

23.2 Make triplicate determinations with all standards and plot the average net absorbance readings against percent TiO_2 on a scale of 1 cm = 0.05% and 1 cm = 0.020 absorbance.

24. Procedure

24.1 Pipet 25 mL of Solution 1 into a 50-mL volumetric flask. Add 5-mL Tiron and 5-mL EDTA, mix, and then add NH_4OH (1:1) dropwise, mixing after each drop, until the color changes through yellow to green, blue, or ruby-red. Then, just restore the yellow color with 1.7 N HCl added dropwise and mixing after each drop. Add 5-mL buffer, dilute to volume, mix, and measure absorbance at 410 nm.

24.2 Blank. Determine blank in the same manner using 25 mL of fusion blank instead of sample.

SPECTROPHOTOMETRIC/EDTA METHODS

25. Calculation

25.1 After correcting for blank, read the percentage of TiO_2 to nearest 0.01 from the calibration curve. Correct for the iron present in the sample to obtain the true TiO_2 as follows: percent $TiO_2 - (0.01 \times$ percent $Fe_2O_3)$.

Phosphorous Pentoxide

26. Reagents

26.1 Molybdic Acid Solution (5.0 N)—Dissolve 25 g of crystalline ammonium molybdate $[(NH_4)_6 Mo_7O_{24} \cdot 4H_2O]$ in about 250-mL water. Carefully add 140-mL H_2SO_4 to 500-mL water while mixing the solution. Cool the solution, and then while mixing, slowly add the ammonium-molybdate solution to the dilute sulfuric acid, and allow to cool. Transfer the cool solution to a 1000-mL volumetric flask, and dilute to volume. Standardize and, if necessary, adjust to 5.0 ± 0.05 N.

26.2 Stock Phosphate Solution (320 ppm P_2O_5)—Dissolve 0.614 g of dried KH_2PO_4 in water, and dilute to 1 L in a volumetric flask.

26.3 Standard Phosphate Solution (32-ppm P_2O_5)—Pipet 50 mL of stock phosphate solution into a 500-mL flask, and dilute to volume.

27. Calibration

27.1 Using 250-mL volumetric flasks, make up solutions containing 0 (blank), 10, 30, and 50 mL of the standard phosphate solution. Add 25 mL 5.5 N HCl to each flask, dilute to volume, and mix. These standards represent 0.0, 0.10, 0.30, and 0.50 percent P_2O_5, respectively.

27.1.1 Pipet 50 mL of each of standard solutions into separate 250-mL beakers and proceed as outlined in Section 28.1 starting with second sentence. Make all tests in triplicate and plot the mean net absorbance against percent P_2O_5 on a scale of 1 cm = 0.10% P_2O_5 and 1 cm = 0.020 absorbance.

28. Procedure

28.1 Pipet 50-mL aliquot of Sample Solution 1 into a 250-mL beaker. With a pipet, add 5 mL of molybdic-acid solution and mix by swirling the beaker. Add 0.2 g of solid ascorbic acid with a measuring scoop delivering 0.2 ± 0.05 g, mix, heat the solution to vigorous boiling, and then boil, uncovered, for 1 min. Cool the solution, transfer it into a 50-mL volumetric flask, and dilute to mark the water. Measure the absorbance at 725 nm.

28.2 Run a blank using 50 mL of the fusion blank in place of the sample, and correct the absorbance reading of the sample accordingly.

29. Calculation

29.1 Read the percentage of P_2O_5 to the nearest 0.01 from the calibration curve.

Manganic Oxide

30. Reagents

30.1 HNO_3 (1:1).
30.2 $NaNO_2$ solution, 5%.
30.3 $AgNO_3$ solution 1%.
30.4 $KMnO_4$ stock solution. Dissolve 0.8009-g $KMnO_4$ in water and dilute to 1000 mL in a volumetric flask.
30.5 Dilute $KMnO_4$ solution—Transfer 100-mL $KMnO_4$ stock solution into a 1000-mL volumetric flask and dilute to mark with water.

31. Calibration

31.1 With a buret, transfer into 250-mL beakers 0.0, 12.5, 25.0, 37.5, and 50 mL of the dilute $KMnO_4$ solution, dilute with water to about 50 mL, add 15-mL HNO_3(1:1) and continue as described in Section 32.1 starting with the second sentence. These solutions represent 0.00, 0.05, 0.10, 0.15, and 0.20% Mn_2O_3. Make triplicate determinations with all mixtures, and plot the mean net absorbance readings against percent Mn_2O_3 on a scale of 1 cm = 0.01 Mn_2O_3, and 1 cm = 0.020 absorbance. Draw a smooth curve through the points.

32. Procedure

32.1 Place a 1.0000-g (Note 12) sample in a 250-mL beaker and disperse with 30 mL of water. Add 15 mL HNO_3 (1:1), and 4 to 5 drops of $NaNO_2$ solution (5%). Boil until the solution is as complete as possible and until the HNO_2 is completely expelled. Remove from

TABLE 2—*Data by Spectrophotometric-EDTA Methods, %.*

	SiO_2	Al_2O_3	Fe_2O_3	CaO	MgO	SO_3	TiO_2	P_2O_5
	\multicolumn{8}{c}{Cement NBS 1014}
	19.60	6.40	2.45	63.00	2.77	2.71	0.26	0.32
	19.50	6.32	2.48	63.15	2.77	2.70	0.24	0.32
	19.60	6.35	2.50	63.11	2.70	2.72	0.25	0.32
Avg	19.57	6.36	2.48	63.09	2.75	2.71	0.25	0.32
Ref[a]	19.49	6.38	2.50	63.36	2.80	2.70	0.25	0.32
	\multicolumn{8}{c}{Cement NBS 1015}
	20.51	4.97	3.27	61.31	4.15	2.24	0.25	0.04
	20.64	4.89	3.32	61.40	4.22	2.26	0.25	0.05
	20.68	5.07	3.30	61.25	4.30	2.25	0.27	0.05
Avg	20.61	4.98	3.30	61.32	4.22	2.25	0.26	0.05
Ref[a]	20.65	5.04	3.27	61.48	4.25	2.28	0.26	0.05
	\multicolumn{8}{c}{Cement NBS 1016}
	20.99	5.03	3.63	65.02	0.38	2.28	0.33	0.13
	20.90	4.91	3.65	65.08	0.38	2.27	0.33	0.14
	21.07	4.87	3.68	65.16	0.38	2.27	0.34	0.13
Ave	20.99	4.94	3.65	65.09	0.38	2.27	0.33	0.13
Ref[a]	21.05	4.97	3.71	65.26	0.42	2.27	0.34	0.13

[a] NBS certificate values.

the hot plate and wash down the sides of the beaker, and dilute with water to about 60 mL. Filter through a Munktells 1F paper into a 250-mL beaker washing the residue and filter paper several times with small portion of water. Keep the total volume of filtrate below 80 mL.

NOTE 12—If the sample contains more than 0.20% Mn_2O_3, take smaller sample size and make the necessary correction in the calculation of the result.

32.1.2 Add 15-mL $AgNO_3$ solution (1%) to the filtrate and stir. Heat to near boiling, add about 5 g of $(NH_4)_2S_2O_8$ and boil 1 min. Cool to room temperature in a cold water bath.

32.1.3 Transfer quantitatively to 100-mL volumetric flask and dilute to volume. Measure the absorbance at 525 nm.

32.2 Run a blank and correct the absorbance reading of the sample accordingly.

33. Calculation

33.1 Read the percent of Mn_2O_3 from the calibration curve.

34. Precision and Accuracy

34.1 Tables 2 and 3 show results using these methods for analysis of NBS SRM standards. The data show that these methods meet the requirements of Table 1 in ASTM Method for Chemical Analysis of Hydraulic Cement (C 114).

TABLE 3—*Data by spectrophotometic-EDTA methods.*

Constituent	SRM No., % by wt							Maximum Difference Allowed[a]
	633	634	635	636	637	638	639	
SiO_2	21.81	20.68	18.53	23.08	22.86	21.47	21.47	...
	21.83	20.72	18.39	23.11	22.92	21.34	21.49	...
avg	21.82	20.70	18.46	23.10	22.89	21.41	21.48	...
ref[b]	21.88	20.73	18.41	23.22	23.07	21.48	21.61	...
Δ_P[c]	0.02	0.04	0.14	0.06	0.06	0.13	0.02	0.16
Δ_A[d]	−0.06	−0.03	+0.05	−0.12	−0.18	−0.07	−0.13	±0.2
Al_2O_3	3.96	5.42	6.26	3.08	3.31	4.45	4.40	...
	3.83	5.27	6.31	3.13	3.42	4.59	4.37	...
avg	3.90	5.35	6.29	3.11	3.37	4.52	4.39	...
ref	3.78	5.21	6.29	3.02	3.28	4.45	4.28	...
Δ_P	0.13	0.15	0.05	0.05	0.11	0.14	0.03	0.20
Δ_A	+0.12	+0.14	0.00	+0.08	+0.09	+0.07	+0.11	+0.2
Fe_2O_3	4.16	2.84	2.60	1.60	1.79	3.53	2.38	...
	4.16	2.82	2.63	1.58	1.78	3.51	2.38	...
avg	4.16	2.83	2.62	1.59	1.79	3.52	2.38	...
ref	4.20	2.84	2.61	1.61	1.80	3.55	2.40	...
Δ_P	0.00	0.02	0.03	0.02	0.01	0.02	0.00	0.10
Δ_A	−0.04	−0.01	+0.01	−0.02	−0.01	−0.03	−0.02	±0.10
CaO	64.70	62.55	59.95	63.61	65.93	62.00	65.67	...
	64.52	62.55	59.87	63.44	66.02	62.20	65.54	...
avg	64.61	62.55	59.91	63.53	65.98	62.10	65.62	...
ref[e]	64.66	62.64	59.95	63.56	66.09	62.16	65.84	...
Δ_P	0.18	0.00	0.08	0.17	0.09	0.20	0.13	0.20
Δ_A	−0.05	−0.09	−0.04	−0.03	−0.11	−0.06	−0.22	±0.3

TABLE 3—Continued

Constituent	SRM No., % by wt							Maximum Difference Allowed[a]
	633	634	635	636	637	638	639	
MgO	1.01	3.34	1.22	4.01	0.81	3.77	1.26	...
	1.07	3.30	1.29	3.98	0.76	3.82	1.39	...
avg	1.04	3.32	1.26	4.00	0.79	3.80	1.33	...
ref	1.04	3.30	1.23	3.95		3.83	1.26	...
Δ_P	0.06	0.04	0.07	0.03	0.05	0.05	0.13	0.16
Δ_A	0.00	+0.02	+0.03	+0.05	+0.02	−0.03	+0.07	±0.2
SO$_3$	2.24	2.23	7.12	2.36	2.41	2.31	2.48	...
	2.23	2.17	7.06	2.34	2.34	2.37	2.42	...
avg	2.24	2.20	7.09	2.35	2.38	2.34	2.45	...
ref	2.20	2.21	7.07	2.31	2.38	2.34	2.48	...
Δ_P	0.01	0.06	0.06	0.02	0.07	0.06	0.06	0.10
Δ_A	+0.04	−0.01	+0.02	+0.04	0.00	0.00	−0.03	±0.1
TiO$_2$	0.24	0.31	0.33	0.18	0.21	0.27	0.33	...
	0.24	0.31	0.31	0.18	0.22	0.25	0.32	...
avg	0.24	0.31	0.32	0.18	0.22	0.26	0.33	...
ref	0.24	0.29	0.32	0.18	0.21	0.25	0.32	...
Δ_P	0.00	0.00	0.02	0.00	0.01	0.02	0.01	0.03
δ_A	0.00	+0.02	0.00	0.00	+0.01	+0.01	+0.01	±0.03
P$_2$O$_5$	0.23	0.09	0.16	0.09	0.22	0.06	0.09	...
	0.24	0.09	0.17	0.09	0.24	0.06	0.09	...
avg	0.24	0.09	0.17	0.09	0.23	0.06	0.09	...
ref	0.24	0.10	0.17	0.08	0.24	0.06	0.08	...
Δ_P	0.01	0.00	0.01	0.00	0.02	0.00	0.00	0.03
Δ_A	0.00	−0.01	0.00	+0.01	−0.01	0.00	+0.01	±0.03
Mn$_2$O$_3$	0.04	0.28	0.09	0.11	0.06	0.05	0.08	...
	0.04	0.29	0.09	0.12	0.06	0.05	0.07	...
avg	0.04	0.29	0.09	0.12	0.06	0.05	0.08	...
ref	0.04	0.28	0.09	0.12	0.06	0.05	0.08	...
Δ_P	0.00	0.01	0.00	0.01	0.00	0.00	0.01	0.03
Δ_A	0.00	+0.01	0.00	0.00	0.00	0.00	0.00	±0.03

[a] Six of the seven values required within limits.
[b] Ref = NBC certificate values for SRM (Standard Reference Materials) portland cements.
[c] Δ_P = measure of precision = difference between individual values of the duplicate tests.
[d] Δ_A = measure of accuracy = difference between duplicate average and the Reference NBS value.
[e] Corrected to include ½ SrO titrated with EDTA.

Section IV
Free Lime Rapid Methods

The present version of ASTM Method for Chemical Analysis of Hydraulic Cement (C 114-85) contains two methods for the determination of free lime in cement and clinker. Method A is a modified Franke method wherein the free lime extracted in an ethylacetoacetate-isobutyl alcohol solvent is titrated with perchloric acid. Method B uses a glycerin-alcohol (ethanol SDA 2B or isobutyl alcohol are specified, but could be replaced by absolute alcohol or ethanol SDA 3A) solvent with a strontium nitrate accelerator to extract free lime, which is then titrated with ammonium acetate. Both of these are rapid methods in which the sample and solvent are boiled for 15 to 20 min, filtered, and titrated. The latter method is a modification of the classical alcohol-glycerol extraction (often referred to as "ASTM," but which last appeared in the 1980 *Annual Book of ASTM Standards*) in which the extraction was boiled for 2 h or longer. Thus, "ASTM" is now much faster than before and is no longer the lengthy procedure that chemists and operators used to curse. References to "ASTM" in the methods that follow refer, of course, to the old method.

The three methods that follow are variations on a theme. They are all based on rapid ethylene glycol extraction with a subsequent titration for free lime as published by Schlapfer and Bukowski in 1933 (see Method 2 for references). It may also interest some to note that current Method A in C 114 is based on the 1941 work of Franke while Method B, modified from the old "ASTM" method, is based on the 1926 publication of Lerch and Bogue.

These methods, as with those in the *Annual Book of ASTM Standards*, do not distinguish between calcium oxide and calcium hydroxide. Consequently, reference to "free lime" is probably correct only when dealing with fresh clinker. In all other cases, "free lime" is undoubtedly a mixture of calcium oxide and hydroxide or calcium hydroxide alone. This has been mentioned many times but is worth repeating again here.

Editor

J. W. Yule[1] and R. D. Chadwick[1]

The Determination of Free CaO in Cements and Clinkers

This method is the fastest of those included. It calls for only a 3-min extraction on a 450°F (232°C) hotplate. It is doubtful that "heating at 450°F (232°C)" actually occurs, however, because the boiling point of ethylene glycol is 390°F (198.83°C from 1984–1985 *Handbook of Chemistry and Physics*). No indication is given if the extraction solution actually boils. Solution temperatures actually obtained would seem to be very dependent on the hotplate used.

Editor

Scope

We have developed and tested a procedure for the determination of free lime in cements and clinkers that is faster than the ASTM procedure (methanol-glycerol solvent) but is as accurate and precise. This procedure uses ethylene glycol to extract the free lime, after which the cement is filtered off and the filtrate is titrated with a standardized solution of hydrochloric acid. It is similar to a procedure found in the International Organization for Standardization (ISO) draft recommendation No. 774 published in 1967 by CEMBUREAU, the European Cement Association, except that our procedure is more rapid and uses a different indicator.

To evaluate the procedure we ran 14 of the most recent interlab cements by the new ethylene glycol procedure and compared them with the interlab averages obtained by the ASTM procedure. The results showed good correlation between the methods as long as the samples were not too old (Table 1). Carbonation of the free calcium oxide (CaO), which occurs gradually with age, caused a lowering of the free lime in the longest stored samples. Only one sample was outside the interlab tolerance and that was one of the older ones.

We have found that heating a 1-g sample in 50 mL of ethylene glycol for 3 min on a 450°F (232°C) hotplate is most suitable for extracting. The heating accelerates extractions and even causes some combined lime to be extracted; however, properly controlled, heating can be used advantageously to speed up the determination without affecting the accuracy of the results. Heating at 450°F (232°C), for example, accelerates the free CaO extraction time about tenfold. For best results, the extraction time and temperature should be held to ±15 s and ±25°F (−4°C). A chart showing the relationship between the extracted lime and the heating time for two different temperatures (450 and 600°F [232 and 316°C]) is shown in Fig. 1.

Because this procedure has not been tested extensively and because of unknown variables that may exist between laboratories, we suggest that each plant lab confirm the procedure

[1] Cement Company, Research Department.

TABLE 1—*Comparison of ethylene glycol procedure with ASTM procedure.*

	ASTM Procedure		New Procedure	
Interlab Number	Interlab Average	Research Rerun	Ethylene Glycol	Repeat
205	0.59	...	0.32	...
206	0.27	0.11	0.12	0.15
207	1.25	...	0.84	...
208	1.25	...	1.12	...
209	1.68	1.28	1.37	1.42
210	0.49	...	0.44	...
211	1.09	...	0.81	...
212	0.72	...	0.57	...
213	0.70	...	0.71	...
214	0.96	0.98	1.11	...
215	0.59	...	0.57	...
216	0.84	...	0.81	...
217	1.29	...	1.27	...
218	1.32	...	1.30	...

by comparing results run by both old and new methods. If there is not good correlation the heating time or temperature could be altered accordingly. Once a desired procedure is established there should be no difficulty in reproducing good results. It is just as important with this new procedure as it is with the ASTM procedure to keep samples, chemicals, and apparatus moisture free and to pulverize clinkers to at least 95 percent passing 100 mesh. Water causes high results, and inadequate grinding causes low results.

This procedure can be performed in 15 to 20 min and is relatively simple. It does require an analyst's nearly constant attention, so performing it becomes less practical the more samples there are to run. It is an excellent procedure for single or dual determinations when speed and accuracy are important. We would appreciate your comments on the suitability of this procedure as a control method. The procedure is as follows.

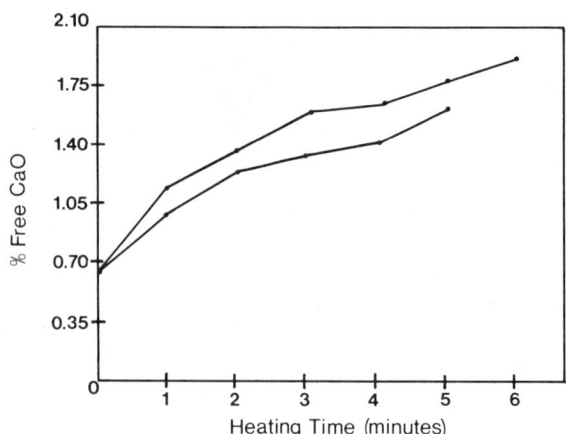

FIG. 1—*Extracted lime versus heating time at 450 and 600°F [232 and 316°C].*

Reagents

- Absolute anhydrous alcohol (reagent alcohol may be substituted).
- Anhydrous ethylene glycol (both of these reagents should be protected from moisture from the atmosphere to insure that they remain anhydrous).
- Minus 50-mesh standard sand (dried at 110°F [43°C]).
- 0.1 N hydrochloric acid.
- Primary standard grade calcium carbonate.
- Phenolphthalein indicator 1.0-g phenolphthalein in 60-mL alcohol diluted to 100-mL with distilled water.

Procedure

Weigh a 1.000-g sample into a dry 250-mL Erlenmeyer flask. Add about 2 g of dry sand and mix thoroughly to break up any lumps of cement. Add 50-mL of ethylene glycol. Heat the flask by placing on a hot plate having a surface temperature of 450°F (232°C) for 3 min. The temperature may be measured with a surface thermometer. Filter the mixture under suction through a medium fritted glass filter into a 125-mL Erlenmeyer flask. The fritted glass filter must be dry. Wash the 250-mL Erlenmeyer flask thoroughly with alcohol and filter through the fritted glass. Wash the residue in the fritted glass three times with alcohol. Add 10 drops of phenolphthalein to filtrate and titrate with 0.1 N hydrochloric acid.

Standardization and Calculations

Place about 0.2000 g of dry, primary standard grade calcium carbonate $(CaCO)_3$ in an ignited and weighed platinum crucible. Ignite the $CaCO_3$ at 1800°F (982°C) for 30 min. Weigh accurately to ±0.1 mg. Treat the resultant CaO as a regular sample.
Calculate

$$F = \text{weight of CaO/mL of } 0.1 \text{ } N \text{ HCl}$$

Calculate percent free CaO in cement of clinker

$$F \times \text{mL of } 0.1 \text{ } N \text{ HCl}$$

Extraction of Free Lime in Portland Cement and Clinker By Ethylene Glycol

Preface

The authors of this ethylene glycol extraction method evaluated the effects of extraction time, extraction temperature, and sample fineness on the results. The method extracts the free lime with ethylene glycol and then titrates the resulting extract with 0.05 N hydrochloric acid. Their basic recommendations are that normal cement fineness (3000 Blaine or higher) is adequate, a temperature range of 80 to 100°C is suitable, as is an extraction time of up to 30 min with a preferred range of 5 to 10 min.

The unique feature of this method is that the glycol is preheated to the desired temperature before adding it to the flask containing the sample. Filtration and washing is done hot as, apparently, is the titration. The titration will be at a temperature below extraction temperature because of the addition of deionized water, presumably at room temperature, immediately before titration.

Editor

Reprinted from *Cement and Concrete Research*, Vol. 12, 1982, pp. 399–403, with permission from Pergamon Press, Ltd., New York.

M. P. Javellana[1] and I. Jawed[1]

Extraction of Free Lime in Portland Cement and Clinker By Ethylene Glycol

ABSTRACT

Free lime in portland cement and clinker has been determined by extraction with hot ethylene glycol. Our determinations show that times of 5-10 minutes and temperatures of 80-100°C are adequate. This method is as accurate as the ASTM method and has the distinct advantage of simplicity and rapidity.

Introduction

Uncombined calcium oxide or "free lime" usually occurs in small amounts (1-2%) in portland cement clinker. The free lime value generally indicates the completeness of the clinkering reactions and the quality of the burning practices. A high free lime clinker usually is due either to inhomogeneity and coarseness of the cement raw mix or to improper burning and cooling conditions in the kiln. For this reason, determination of free lime in portland cement clinker, together with some other physical characteristics, provide a convenient and useful method for assessing clinker quality.

The most widely used methods for the determination of free CaO in cement and cement clinker are based on its extraction in an organic solvent. The ASTM recommended methods are based on the work of Lerch and Bogue (1), which uses the solubility of CaO in a glycerol-absolute alcohol solvent, and that of Franke (2,3), which uses CaO dissolution in acetoacetic ester-isobutanol solvent. However, these methods are tedious and time-consuming.

A simple and relatively rapid method, described by Schlapfer and Bukowski (4,5), is based on the extraction of CaO by ethylene glycol heated to 60-70°C. This method gives results in reasonable agreement with those obtained by the glycerol-alcohol method and reduces the extraction time to about half an hour. Recently, Wang et al (6) used the same method to show that ethylene glycol at 80-85°C reduced extraction time to less than 5 minutes and still provided results that agreed well with those obtained by other methods. In view of this information, we have evaluated the ethylene glycol procedure in some detail as a function of time, solvent temperature, and cement clinker fineness.

It is not our purpose to discuss the chemistry of CaO dissolution in organic solvents. This subject has been discussed in detail by Swenson and Thorvaldsen (7) and Longuet et al (8). The present work stresses the experimental aspects of CaO extraction by ethylene glycol with the objective of establishing a rapid and reliable method for determination of free lime in cement and cement clinker for possible application to quality control problems in cement production.

[1] Martin Marietta Laboratories, Baltimore, MD.

Experimental

Equipment

Burette (0.5-ml subdivision)
Erlenmeyer flask with rubber stopper (250 ml)
Hot plate with stirrer
Thermometer
Teflon-coated magnetic stirring bar
Vacuum filtering flask (250 ml)
Gooch crucible with holder
Glass fiber filter

Reagents

Anhydrous ethylene glycol
0.05 N hydrochloric acid (standardized against 0.05 N sodium hydroxide which is standardized with potassium acid phthalate)
Phenolphthalein indicator (1% solution in ethanol)

Materials

Several cement and clinker samples were used for this evaluation. The cement samples were supplied by the Cement and Concrete Research Laboratory of the National Bureau of Standards as part of the CCRL test. The clinkers were obtained from Martin Marietta cement plants.

Procedure

A sample (usually 1 g) of ground cement or clinker was accurately weighed into a clean, dry, 250-ml Erlenmayer flask. It is important that the sample does not come in contact with moisture before extraction. About 50 ml of ethylene glycol, heated to the desired temperature (50-100°C), were added to the flask and stirred for the desired amount of time (5 min to 3 hr) using a magnetic stirrer on a hot plate. The temperature was maintained at the desired value and checked frequently. The hot suspension was filtered under suction through a glass fiber filter fitted onto a Gooch crucible and the clean filtrate was collected in a 250-ml vacuum filter flask. The filter should be wetted with ethylene glycol before filtering. The Erlenmeyer flask was rinsed at least twice with about 10-15 ml hot ethylene glycol which, after filtering, was added to the original filtrate.

About 25 ml of deionized water and 1 ml of phenolphthalein indicator solution were added to the filtrate and then titrated to a colorless endpoint with 0.05 N HCl. Addition of water to the ethylene glycol filtrate facilitates the endpoint determination. Free lime is calculated as follows:

$$\% \text{ Free CaO} = \frac{\text{ml HCl} \times \text{normality of HCl}}{10 \times \text{sample weight}} \times 28$$

For comparison the free lime was also determined by the ASTM method using the glycerol-ethanol solvent (9).

Results

Comparison of Methods

The results of the free-lime determination by the ethylene glycol and ASTM methods are given in Table 1.

TABLE 1

Comparison of Methods

Sample*	Free CaO (%)	
	Ethylene Glycol	ASTM
CCRL #53	0.61	0.63
CCRL #55	0.81	0.83
CCRL #57	0.56	0.57
CCRL #59	0.77	0.75
CCRL #60	0.54	0.53
CCRL #61	1.18	1.25
CCRL #62	1.36	1.36

*As-received cements (Blaine: approx. 3000 cm^2/g) extracted for 5 min at 80-85°C. Excellent agreement is shown between the two methods.

Effect of Extraction Time

To determine the effect of extraction time on CaO dissolution, extractions for periods of time ranging from 5 minutes to 3 hours were done. The results are shown in Table 2.

TABLE 2

CaO Dissolution in Ethylene Glycol with Time

Sample*	Extraction Time	Free CaO (%)
CCRL #53	5 min	0.61
Free CaO by	10 min	0.65
ASTM Method:	30 min	0.69
0.63%	1 hr	0.81
	2 hr	0.90
	3 hr	0.98
CCRL #61	5 min	1.18
Free CaO by	10 min	1.18
ASTM Method:	30 min	1.27
1.25%	1 hr	1.51
	2 hr	1.55
	3 hr	1.54

*As-received cements (Blaine: 3000 cm^2/g) extracted at 80-85°C.

With time, more CaO is extracted. This result could be due to the dissolution of other calcium-containing clinker minerals after prolonged extraction, e.g. calcium silicates and aluminates, and gypsum. However, times up to 30 minutes do not seem critical.

Effect of Extraction Temperature

The temperature of the ethylene glycol was varied from 50°C to 100°C to determine the effect of solvent temperature on CaO dissolution. The results are given in Table 3.

TABLE 3
Effect of Ethylene Glycol Temperature on CaO Dissolution

Sample*	Temperature of Ethylene Glycol (°C)	Free CaO (%)
CCRL #60	50	0.41
Free CaO by	60	0.45
ASTM Method:	80	0.54
0.53%	100	0.54
CCRL #61	50	1.01
Free CaO by	60	1.13
ASTM Method:	80	1.18
1.25%	100	1.17

*As-received cement (Blaine: approx. 3000 cm^2/g) extracted for 5 min.

With increase in solvent temperature, the CaO dissolution seems to increase, reaching a near constant value at about 80°C. A temperature range of 80-100°C seems suitable for extraction.

Effect of Sample Fineness

Clinkers from two different cement plants were ground to two different finenesses to assess the effect of sample fineness on the CaO dissolution. The results are given in Table 4.

TABLE 4
Effect of Sample Fineness on CaO Extraction

Sample*	Fineness (cm^2/g Blaine)	Free CaO (%) Ethylene Glycol	ASTM
Clinker A	2950	1.89	1.96
	5930	1.98	2.02
Clinker B	3110	1.57	1.58
	6100	1.46	1.38

*Clinker extracted for 5 min at 80-85°C

Increase in clinker fineness above ~3000 cm^2/g Blaine does not seem to show any effect on free CaO extraction within experimental error.

Conclusions

Our work shows that hot ethylene glycol extraction offers a very rapid and convenient method for determination of free CaO in fresh cement and clinkers. Extraction times of 5-10 minutes and temperatures of 80-100°C are quite adequate for this determination. Although this method is as good as the ASTM method for accuracy, its simplicity and rapidity give it a distinct advantage. It is most suitable for quality control during cement clinker production, where time can be an important factor. It is not suitable for cement and clinker stored for long periods of time or in contact with moisture since ethylene glycol would dissolve both CaO and Ca(OH)$_2$. However, the ASTM method suffers from the same shortcomings. In our opinion, the method deserves serious consideration by the ASTM committee on portland cement analysis as an alternate method for free-lime determination in portland cement and clinker.

Acknowledgment

This work was supported by the National Science Foundation through grant no. CME-79-02665. Thanks are also due to A. Kolan for help with this work.

References

1. W. Lerch and R. Bogue, Ind. Eng. Chem., 18, 739 (1926).

2. B. Franke, Z. anorg. allg. Chem., 247, 180 (1941).

3. E. Presler, S. Brunauer and D. Kantro, Anal. Chem., 28, 896 (1956); 33, 877 (1961).

4. P. Schlapfer and R. Bukowski, Rep. Swiss Fed. Lab. Test Mater., Zurich, No. 63 (1933).

5. P. Schlapfer, Proc. Intl. Symp. Chem. Cement. Stockholm, p. 289 (1938).

6. C. Wang, C. Chang, C. Chen and W. Cheng, Rev. 26th Gen. Mtg. Cem. Assoc., Japan, p. 76 (1972).

7. E. Swenson and T. Thorvaldson, Can. J. Chem., 29, 140 (1951); 30, 257 (1952).

8. P. Longuet, L. Burglen and G. Belline, C.E.R.I.L.H. Sp. Rep. (1977).

9. Annual Book of ASTM Standards, Part 13, 134 (1981), Am. Soc. Test. Mater., Philadelphia, Pa.

APPENDIX

This is an independent evaluation of the preceding method. It verifies agreement of the method with the classical glycerin-ethanol method that used to be in ASTM C 114.

Editor

Free Lime Determination by Ethylene Glycol Extraction By N. T. Flores[2]

Two cement samples from R. Pyrdeck were tested for free CaO using the ethylene glycol method as described in the previous paper, as well as the British method (Table 1). The value by the ASTM method was from the Control Laboratory. The extraction time used was 10 min at 80 to 90°C. The data shows that the ethylene glycol value is closer to the value by ASTM as compared to the British method. The method is very easy, reproducible, and not time consuming.

TABLE 1—*Free lime determination (British and ASTM methods versus ethylene glycol method).*

Method	%Free CaO		
	ASTM	Ethylene Glycol	British
Sample 76416	1.53	1-1.56	1.80
		2-1.58	1.82
		3-1.59	
		4-1.59	
		5-1.61	
Sample 76426	0.32	1-0.36	0.51
		2-0.36	0.51

REQUIREMENTS: 0.0496 N HCl; ethylene glycol (Anhydrous); phenolpthalein—0.5% in anhydrous and neutralized to faint pink with NaOH dissolved in anhydrous alcohol.

[2] California Portland Cement Company, Colton, CA.

LaFarge Chemical Method No. 42: Free CaO[1]

Introduction

This method specifies a longer time than the other methods, including those now in ASTM Method for Chemical Analysis of Hydraulic Cement (C 114). It has two unique features in that the heating of the extraction involves 30 min in a 110°C oven and then, after suitable preparation steps, determines actual CaO extracted by means of a complexometric titration using ethylenediaminetetraacetate (EDTA) at high pH.

Editor

Reagents, Equipment, and Supplies

- Ethylene glycol, anhydrous
- Absolute ethyl alcohol
- EDTA solution, standardized against CaO
- 2 N NaOH solution
- Triethanolamine: water, 1:2 by volume
- c.HCL:water, 1:1 by volume
- Chrome blue-black indicator, 2 g in solution of 50-mL triethanolamine + 50-mL water
- pH indicator paper, 11.0 to 14.0
- 110°C oven
- Vacuum pump (water) with check valve in line
- 500-cm^3 vacuum Erlenmeyer flask
- Fritted disk Pyrex Bucher funnel, fine, 65-mm diameter (Fisher Catalogue No. 10-358K)
- Magnetic stirrer
- Pyrex beakers, 100-mL capacity, with watch glass
- Plastic wash bottle for alcohol, 100 to 200 mL capacity

Table 1 illustrates the LaFarge Chemical Method No. 42 for Free calcium oxide (CaO).

[1] Canada Cement LaFarge Ltd., 506 Cathcart, Montreal 111, Quebec, Canada.

FREE LIME RAPID METHODS 85

TABLE 1—*Application of complexometry to determination of free lime in cement and clinker after dissolution in glycol.*

	Steps	Key Points	Explanation
1	putting free CaO into solution	material ground finer than 100-μm (No. 140) sieve	to help dissolution
		2-g material 100-mL beaker, dry	water liberates CaO from silicate causing high result for lime
		25-mL ethylene glycol, mix cover with small watch glass 30 min in 110°C oven	to condense glycol vapor necessary to dissolve all the free lime to help clean beaker and to get better filtration
2	filtering the solution	before filtering, shake if necessary to suspend material stuck to bottom of beaker. Add a little alcohol to rinse if necessary; dry 500-mL conical suction flask, with check valve filter through Pyrex fritted fine Bucher funnel (Fisher 10-358K) and rinse with alcohol; dry filter all solution under vacuum, rinse beaker and funnel with about 40-mL alcohol repeat further two times filtrate must be completely clear	to avoid water running back into flask no water in contact with material no water in contact with material if particles pass into filtrate free CaO result will be high fine fritted funnel retains very fine particles
		after using once or twice, clean the funnel with acid to the flask containing filtrate, add several drops 1:1 HCL and 300-mL distilled water	to avoid hydrolysis of CaO on adding water
3	preparing solution for pH test	conical flask placed on magnetic stirrer, no heating add 15-mL 1:2 triethanolamine from very clean test tube or other container add 40-mL 2 N NaOH from very clean container, to give pH of 12.5 to 13.5 check pH with pH paper	better observation of color change if solution is dilute to complex Fe_2O_3 and Al_2O_3 possibly present, which could interfere with CaO determination if pH is less than 12.5 CaO result will be high, because of complexing of MgO if pH is above 13.5, color change will not be as clear

TABLE 1—Continued

4	determination of CaO	indicator chrome blue-black about 8 to 9 drops added to give red-violet color	red-violet color obtained in presence of free Ca ions
		titrate with EDTA solution through violet to a clear blue color	clear blue coloration indicates disappearance of calcium ions
		read mL of EDTA used	
		add 3 or 4 drops more of EDTA solution if blue color has not changed, first reading of mL EDTA is final reading; otherwise record reading for final addition when blue color is stable if end point is not clear, check pH again	at least 3 drops are necessary to cause a change in color
		N mL of EDTA solution used for titration	1-mL EDTA equivalent to t grams CaO
5	calculation of free CaO	t is EDTA factor $$\% \text{ CaO} = \frac{Nt \times 100}{2}$$	2-g material used for determination

NOTE: For very high or very low concentrations of free CaO, sample weight of material may be varied; for w grams sample, $\% \text{ CaO} = Nt \times 100/w$.

Section V
Rapid Analysis of Sulfur

  The only method included is an instrumental method using the LECO induction furnace and automatic titrator. As noted by the authors, it is essentially the method given in a LECO Applications Bulletin with changes in sample size and the combination of accelerators. It determines total sulfur as sulfur trioxide (SO_3).
  Data included at the end of the method show qualification under the requirements of ASTM Method for Chemical Analysis of Hydraulic Cement (C 114). Reproducibility is excellent for all samples, and only for the high sulfur Type K SRM does the average of two runs approach the allowable difference from the SRM value.

Editor

M. G. Lewis[1] *and E. H. Scott*[1]

Method for Analysis of Total Sulfur as SO_3 in Portland Cement and Clinker Using the LECO Sulfur Analyzer[2]

1. Scope

1.1 This method covers the sulfur trioxide (SO_3) analysis of portland cements and clinker in the 0.1 to 4.5% range. A sample is combusted in an induction furnace so that the sulfur is converted to sulfur dioxide (SO_2) and carried over to a titrating vessel. The vessel contains free I_2 and starch, which gives the solution a blue color. As the SO_2 reduces the I_2 and ties it up as HI, the solution becomes colorless. During the reduction of I_2, the automatic titrator adds more free I_2, in the form of potassium iodate (KIO_3), to restore and maintain the original blue color. From the amount of KIO_3 added, the $\%SO_3$ can be calculated.

2. Limitations

2.1 This method was developed primarily for portland cement and clinker but can be expanded to include other materials such as cement dust and raw mix.

3. Apparatus and Materials

3.1 Induction furnace, LECO Model 521 or equivalent.
3.2 Automatic titrator, LECO Model 532-000 or equivalent.
3.3 Oxygen cylinder equipped with two stage regulator.
3.4 Purifying train (for oxygen), LECO Model 516-000 or equivalent.
3.5 Variable transformer, LECO Model 521-084 or equivalent.
3.6 Timer, LECO Model 593-100 or equivalent.
3.7 Crucible, LECO #528-035 or equivalent.
3.8 Crucible cover, LECO #528-042 or equivalent.
3.9 Combustion tubes, LECO #550-120 or equivalent.
3.10 Accelerator scoop, LECO #503-032 or equivalent.

4. Reagents

4.1 KI crystals.
4.2 KIO_3 solution, 0.8436 g/L.

---

[1] Martin Marietta Cement Technical Center, 1450 S. Rolling Rd., Baltimore, MD 21227.
[2] Method is similar to method for cement and clinker published in LECO Applications Bulletin. Method differs in sample size and combination of accelerators. LECO method prescribes 0.1-g sample with 1-g iron chips, ½ copper ring, and 1½ scoops of tin metal. This method prescribes 0.2-g sample with one copper ring, one scoop of iron powder, and two scoops of tin metal.

4.3 Arrowroot Starch Solution. Mix 10 g of arrowroot starch with 50-mL distilled water. Separately boil 250-mL distilled water. Add the 250 mL of boiling water to the 50 mL of water containing the starch solution and mix thoroughly. Mix in 12 mL of 30% sodium hydroxide (NaOH) solution. Dilute to 1 L. Add 24 g of KI.
4.4 Iron powder accelerator, LECO #501-078 or equivalent.
4.5 Copper ring accelerator, LECO #550-184 or equivalent.
4.6 Tin metal accelerator, LECO #501-076 or equivalent.
4.7 Glass wool.
4.8 NaOH pellets.
4.9 National Bureau of Standards (NBS) cement standard reference material (SRM) series 633 through 639.
4.10 HCl solution 15-mL concentrated HCl/L.

5. Procedure

5.1 Weigh 0.2000 g of an NBS cement SRM (or suitable level of SO_3) to four places on an analytical balance into a crucible. Record weight of NBS cement. Add the following accelerators: one copper ring, one scoop of iron powder, and two scoops of tin. Shake gently. After adding the accelerators place a cover on the crucible. Prepare two of the NBS cement standard. Prepare cement sample in same manner.

5.2 Turn on the high voltage switch and the filament switch on the furnace. Turn on the power switch #532-31 in the neutral (middle) position on the titrator. Set grid current tap switch on furnace to the middle position.

5.3 Open the stopcock below the titrating vessel, and rinse the vessel with 15% HCl solution several times. Close the stopcock and add 15% HCl solution until level is 3/4 to 1 in. (19 to 25.4 mm) above the bottom of the titrating vessel. Turn on oxygen flow so that purifying train is indicating 1 L/min. Add 5 mL of the starch solution to the titrating vessel. Fill the titrating buret with the KIO_3 solution by depressing the titrator manifold while squeezing the rubber bulb until the KIO_3 solution fills the buret.

5.4 Turn the titrator end point adjust switch #549-26 as far to the left (counterclockwise) as it will go. Flip double throw switch #532-31 down to end point position. Slowly rotate the end point adjust switch #549-26 to the right (clockwise) until the solution in the titrating vessel turns medium to dark blue in color. It will require approximately 0.006 buret units. There should be no need to readjust the end point unless for some reason the instrument is accidentally put out of adjustment. Flip titrator double throw switch #532-31 back to neutral position. Refill the titrating buret with KIO_3 as before.

5.5 Place the crucible containing the NBS cement[3] and accelerators on the pedestal in the loading apparatus and elevate it into the furnace. Flip double throw switch #532-31 to the titrate position. Set the transformer on 88.[4] Check to determine if oxygen flow is 1 L/min and readjust if necessary. Set the timer for 5 min. Depress button on timer to start furnace. When the timer stops the furnace, flip double throw switch #532-31 to the neutral position. Read titrating buret and record reading. Lower the crucible from the furnace and remove with tongs. Repeat procedure with second standard. Take average of two determinations. The two determinations of each NBS cement standard should agree within 0.05% SO_3 or better. Refill buret in between each titration.

5.6 Repeat titration procedure for cement sample in duplicate and take average of readings.

[3] Before running NBS cements standard and cement samples, one or two blanks should be run with crucibles containing accelerators only.
[4] Transformer setting may vary for different laboratories depending on the line voltage coming into the laboratories.

TABLE 1—*Qualification data.*

NBS SRM Number	NBS Value	Run 1	Run 2	Difference Between Runs	Specified Difference Allowed Between	Two Runs	Difference of Average From SRM	Specified Difference Allowed
633	2.20	2.25	2.26	0.01	0.10	2.25	0.05	±0.1
634	2.21	2.17	2.15	0.02		2.16	0.05	
635	7.07	6.95	6.97	0.02		6.96	0.11	
636	2.31	2.31	2.28	0.03		2.30	0.01	
637	2.38	2.33	2.33	0.00		2.33	0.05	
638	2.35	2.32	2.32	0.00		2.32	0.03	
639	2.48	2.45	2.42	0.03		2.43	0.05	

6. Calculation

$$\%SO_3 = \left[\frac{\left(\begin{array}{c}\text{sample}\\ \text{buret reading}\end{array}\right) - \left(\begin{array}{c}\text{average blank}\\ \text{buret reading}\end{array}\right)}{\text{sample weight}}\right] \times \text{conversion factor}$$

where conversion factor = 5.92×0.8436 g KIO_3 = 4.994.

7. Qualification Data

Table 1 shows the qualification data.

APPENDIX I
Alumina: Direct Determination Chemical Method

Over many years, many analysts and ASTM members have expressed the opinion that ASTM Methods for Chemical Analysis of Hydraulic Cement (C 114) needs a direct method for determining alumina (Al_2O_3) rather than determining it by difference as in the present scheme of chemical analysis. This "by difference" approach has generated considerable controversy over the years as to just what was to be subtracted from the Ammonium Hydroxide Group (R_2O_3) to determine "Al_2O_3" and what was to be used in calculating the Bogue potential compounds both for research and for specification purposes.

To some degree, the controversy has been somewhat laid to rest in that C 114 now says, "subtract the phosphorus pentoxide (P_2O_5) and titanium dioxide (TiO_2) as well as the ferric oxide (Fe_2O_3) when determining Al_2O_3," while ASTM Specification for Portland Cement (C 150) now says, "add the P_2O_5 and TiO_2 to the Al_2O_3 determined to get 'Al_2O_3' for purposes of calculating compounds." This, plus the widespread and growing use of instrumental methods, has largely obviated the need for a direct method for alumina in the specific methods sections of C 114. In some cases, however, need for such a method may exist.

In some old research files at Lehigh Cement Company Headquarters, a manuscript was located of work dated Nov. 1951, by C. L. Ford, of the Portland Cement Association, that covers exactly the subject of interest. Further, it apparently was intended for submittal to ASTM for consideration and represents some very thorough work.

In any case, here it is for the benefit of anyone who would like to make use of it.

Editor

C. L. Ford[1]

A Chemical Method for the Direct Determination of Aluminum Oxide in Portland Cement

Synopsis

A direct method for the determination of aluminum oxide in portland cement is described. By using the precipitate of the ammonium hydroxide group for the determination, the method can be included as a part of the ASTM scheme of analysis of portland cement.

The ammonium hydroxide precipitate is ignited, fused with potassium bisulfate, and the melt dissolved in dilute sulfuric acid. Iron and titanium are precipitated with a sodium hydroxide solution and removed by filtration. The filtrate is acidified with hydrochloric acid, nearly neutralized with ammonium hydroxide, and the aluminum precipitated with 8-hydroxyquinoline in an acetic acid-acetate solution. The aluminum oxide content of the precipitate can be determined by (1) weighing the precipitate after drying to constant weight at 120 to 140°C, (2) dissolving the precipitate in hydrochloric acid and titrating with a standard solution of potassium bromate and potassium bromide, or (3) decomposing the precipitate with nitric and sulfuric acids, precipitating the aluminum as hydroxide, igniting the precipitate and weighing as aluminum oxide and correcting for silica.

Introduction

ASTM Specifications for Portland Cement (C 150-49) and those for Air-Entraining Portland Cement (C 175-48T) contain a maximum limit for aluminum oxide (Al_2O_3) and tricalcium aluminate (C_3A) content for some types of cements. In the course of the usual analysis of cement, the Al_2O_3 (and calculated C_3A) content is determined (except when a sample fails to meet specification requirements) by subtracting the separately determined amount of ferric oxide (Fe_2O_3) from the total amount of oxides precipitated by ammonium hydroxide (NH_4OH), the difference being assumed to be entirely Al_2O_3. The values thus calculated are too high because phosphorus pentoxide (P_2O_5) and titanium dioxide (TiO_2), seldom absent in portland cement, are precipitated by NH_4OH along with the Al_2O_3 and Fe_2O_3. In the author's experience the positive error caused by P_2O_5 and TiO_2 usually ranges from 0.25 to 0.40% although at times it may be as high as 0.75%. Hence, when more accurate Al_2O_3 values are needed to meet specification requirements or for other purposes, the interference caused by P_2O_5 and TiO_2 must be eliminated. This may be done by either (1) determining these components separately and calculating the Al_2O_3 content by difference or (2) determining the Al_2O_3 content directly by procedures that eliminate the interfering components.

[1] Chief, Analytical Laboratories, Research and Development Division, Portland Cement Association, Chicago, IL.

According to the current ASTM Methods for Chemical Analysis of Hydraulic Cements (C 114), aluminum oxide is determined by difference. The precipitate of the ammonium hydroxide group (commonly referred to as R_2O_3) is ignited, and the weight of the ignited precipitate determined. Ferric oxide (Fe_2O_3), phosphorus pentoxide (P_2O_5), and titanium dioxide (TiO_2) are then determined on separate samples and subtracted from the weight of the ignited precipitate to obtain the aluminum oxide. This procedure is rather lengthy, and the final result obtained for aluminum oxide is subject to the summation of possible errors involved in the determination of the total oxides and the oxides of iron, phosphorus, and titanium. Further, the current ASTM specifications for cement contain no limitations on P_2O_5 and TiO_2; hence a direct method for Al_2O_3 would save time by making unnecessary their determination for acceptance tests.

A polarographic method for the direct determination of aluminum oxide has been described by Ford and LeMar [1]. Although the method shows considerable promise with respect to accuracy, it requires costly equipment and personnel with special training.

The purpose of the present paper is to describe an accurate and rapid direct chemical method, which has been developed in these laboratories.

Preliminary Studies

Since current ASTM methods already provide for the precipitation and separation of the ammonium hydroxide group, this study has been confined to a search for methods to separate aluminum oxide from the other constituents precipitated by ammonium hydroxide, which, in the case of portland cement, are those named above. Numerous procedures are described in the literature. Of these, at least two [2,3] were developed specifically for cement and are described briefly below.

Chandler [2] proposed a method by which the cement is dissolved in dilute hydrochloric acid and the iron and titanium separated by precipitation with cupferron. The aluminum in the filtrate is then precipitated with 8-hydroxyquinoline. The aluminum oxide content of the precipitate is determined either gravimetrically by weighing the dried aluminum oxyquinolate or by titration with a standard solution of potassium bromate and potassium bromide.

According to the second method developed by Kampf [3], silica is removed, then the aluminum and iron in the filtrate are precipitated together with 8-hydroxyquinoline. The precipitate, after dissolving in acid, is titrated with a standard solution of potassium bromate-bromide, using methyl red as an internal indicator. The iron oxide content is determined on the same sample by titration with a standard sodium thiosulfate solution, and the equivalent value subtracted from the total potassium bromate-bromide titration. Comparative laboratory tests of Chandler's and Kampf's methods gave rather unsatisfactory results. They will be discussed in more detail later.

Of all the other procedures studied, none was found that separated all the interfering elements in one operation, hence a study was made of stepwise separations. Methods for separating aluminum from iron have been based on (1) keeping the iron in solution while precipitating the aluminum and (2) precipitating the iron and keeping the aluminum in solution.

Iron has been kept in solution by reduction to the ferrous state only [4–6] and by reduction to the ferrous state followed by the formation of soluble iron complexes [7–9]. Our results obtained by reducing iron to the ferrous state alone were unsatisfactory. Of the procedures utilizing the formation of iron complexes, only one, using thioglycolic acid as suggested by Welcher [8], was tried. The method was fairly successful, but the offensive odor of the reagent made a search for other methods desirable. The Smith et al. [9] procedure using 2,2'-bipyridine, and the Kassner and Ozier [7] procedure using potassium cyanide were not tested because of the high cost of 2,2'-bipyridine and the poisonous nature of potassium cyanide.

A number of methods for precipitating and separating iron (and in some cases titanium) from aluminum have been developed. Phenylhydrazine [4], urea and succinic acid [5,6], ammonium carbonate [9], cupferron [2,4], and sodium hydroxide [10–12] solutions have been used as precipitating agents. The last two, cupferron and sodium hydroxide, are more useful than the others because they remove titanium as well as iron. The procedure using cupferron has disadvantages in that the time-consuming process of destroying the excess of the cupferron reagent in the filtrate must be completed before the aluminum can be precipitated and that the cupferron precipitate is often very bulky and may adsorb some of the aluminum, thus giving low results. The procedure using sodium hydroxide not only gives quantitative separations, but provides a solution that requires only neutralization before the separation of aluminum from phosphorus. Comparative studies on the two procedures in these laboratories indicated the latter method to be preferable.

The size of the sodium hydroxide precipitate depends on the amount of iron present in the sample analyzed. Since all precipitates have a tendency to adsorb other substances, a study was made to determine the amount of aluminum retained in the sodium hydroxide precipitate. Bright and Fowler [11] in tests conducted on the separation of aluminum from iron and nickel by precipitation with sodium hydroxide, found that in a solution containing 0.0400-g aluminum, 0.040-g nickel, and 0.220-g iron, the loss of aluminum in the sodium hydroxide precipitate, after a single precipitation, was 0.0005 g. In another solution containing 0.0400-g aluminum, 0.004-g nickel, 0.050-g chromium, and 0.240-g iron, the loss was 0.0004 g. However, in each of their two solutions, the amount of iron present far exceeds the iron content of any cement sample used in this work, and furthermore, nickel is not present in cement. Our own findings, based on a study of the five cement samples used for testing, indicated the loss to be negligible.

According to the literature [10–17] 8-hydroxyquinoline has been used extensively to quantitatively precipitate aluminum as the oxyquinolate after the separation of iron and titanium, leaving phosphorus in solution. Lundell and Knowles [13] have shown that after a single precipitation in a solution containing 0.05 g of aluminum, none could be found in the filtrate. In another solution containing 0.05 g of aluminum and 1.0 g of phosphorus pentoxide, the precipitate contained less than 0.0002 g of phosphorus pentoxide. In establishing a suitable pH for the quantitative precipitation of aluminum, Knowles [16], basing his conclusion on the investigations of Fleck and Ward [14] and Goto [15], states that the pH should never be less than 4.2 nor in excess of 7.0. In our work, described later, precipitation was conducted at a pH of approximately 5.0.

The following three procedures have been reported for determining the aluminum content of the oxyquinolate precipitate:

The precipitate is (1) dried to constant weight at 120 to 140°C and weighed as the anhydrous compound [7,11,13,16], (2) dissolved in hydrochloric acid and the solution thus obtained titrated with a standard solution of potassium bromate-bromide [11,13,16], and (3) submitted to wet oxidation to destroy the organic matter; then the aluminum is determined by precipitating with ammonium hydroxide and igniting and weighing as the oxide [13,16]. Knowles [16] has shown that if the amount of aluminum does not exceed 50 mg, good results may be obtained by weighing the dried aluminum oxyquinolate, that amounts greatly exceeding 25 mg should not be titrated, and that amounts exceeding 50 mg should be determined by weighing as the oxide.

The studies and preliminary tests discussed above led to the development of a general procedure for the separation of aluminum as the oxyquinolate and three possible methods of determining the aluminum content of the aluminum oxyquinolate. The detailed procedure is given in Appendix I. An outline of the procedure is presented below.

Procedure

Silica is removed (but not weighed) according to ASTM C 114-47, Section 33(a), 8(c), and 8(d) through the stage where the silica is volatilized and the residue ignited. The ammonium hydroxide group is precipitated according to Section 9(a) and 9(b) and ignited in the crucible containing the residue from the silica separation.

The ignited precipitate is fused with a small amount of potassium bisulfate and dissolved in dilute sulfuric acid. Iron and titanium are precipitated with a sodium hydroxide solution and removed by filtration. The filtrate containing both aluminum and phosphorus is acidified with hydrochloric acid and nearly neutralized with ammonium hydroxide. Precipitation of the aluminum with 8-hydroxyquinoline is conducted in an acetic acid-acetate solution. The reaction takes place according to the following equation

$$AlCl_3 + 3\ C_9H_6NOH \longrightarrow Al(C_9H_6NO)_3 + 3HCl \tag{1}$$

After the aluminum has been precipitated as aluminum oxyquinolate, the aluminum oxide content of the precipitate can be determined in one of three ways:

1. The precipitate may be dried to constant weight at 120 to 140°C, cooled, weighed, and the aluminum oxide content calculated.

2. The precipitate may be dissolved in hydrochloric acid and titrated with an excess of a standard solution of potassium bromate-bromide. The aluminum oxyquinolate is quantitatively bromated to 5,7-dibromo-8-hydroxyquinoline according to the following equations

$$Al(C_9H_6NO)_3 + 3HCl \longrightarrow AlCl_3 + 3C_9H_6NOH \tag{2}$$

$$KBrO_3 + 5KBr + 6HCl \longrightarrow 3Br_2 + 6KCl + 3H_2O \tag{3}$$

$$C_9H_6NOH + 2Br_2 \longrightarrow C_9H_4Br_2NOH + 2HBr \tag{4}$$

Potassium iodide is added to the solution containing excess bromine liberated according to Eq 3 liberating free iodine

$$2KI + Br_2 \longrightarrow 2KBr + I_2 \tag{5}$$

The iodine liberated according to Eq 5, which is proportional to the excess potassium bromate-bromide solution, is then titrated with standard thiosulfate solution using starch as an indicator

$$2Na_2S_2O_3 + I_2 \longrightarrow 2NaI + Na_2S_4O_6 \tag{6}$$

3. The precipitate may be decomposed with nitric and sulfuric acids, the aluminum precipitated as hydroxide, ignited to the oxide, and corrected for silica.

Experimental Data and Discussion of Results

1. Determination of Aluminum Oxide in a Synthetic Solution

Preliminary tests were made on the reliability of the proposed method using a synthetic solution containing a known content of aluminum. In order to simulate a solution of the oxides precipitated by ammonium hydroxide in the usual analysis of 0.5 g of portland cement, known amounts of iron, phosphorus, and titanium were added to the solution.

Aluminum wire of 99.99% purity was used as a source of aluminum, iron wire of 99.89% purity for iron, C. P. KH_2PO_4 for phosphorus, and NBS Sample 154 for titanium.

A 25-mL aliquot contained 0.0300-g aluminum calculated as Al_2O_3, 0.0200-g iron as Fe_2O_3, 0.0030-g phosphorus as P_2O_5, and 0.0020-g titanium as TiO_2.

The solution was analyzed for "R_2O_3," Fe_2O_3, P_2O_5, and TiO_2. The value for Al_2O_3 was calculated by difference. The analyses were made in accordance with the ASTM procedures for portland cement shown in Table 1.

The results of the analysis of this solution by ASTM procedures are shown in Table 2.

The values for Al_2O_3 calculated "by difference" show good agreement with the actual value. The greatest deviation from the actual value is 0.0004 g, the smallest is 0.0001 g.

The synthetic solution was then analyzed by the direct method, using all three ways for the determination of the aluminum oxide content of the oxyquinolate precipitate. These results also are given in Table 2. A comparison of the values shows that (1) experimental results are in very good agreement with the calculated ones, (2) agreement between the three methods is good, and (3) when there is a difference between two results, the difference is within the limits permitted by the ASTM C 114.

2. Determination of Aluminum Oxide in Portland Cements by ASTM Procedures

The value of the method for the analysis of a solution having been demonstrated, the practical applicability of the method was tested by determining the aluminum oxide contents of five different portland cements. The cements were selected for the reasons shown in Table 3.

The "R_2O_3," Fe_2O_3, P_2O_5, TiO_2, and Al_2O_3 were determined by the ASTM procedures named above. SiO_2 (present as an impurity in the "R_2O_3") was determined according to ASTM C 114-47, Section 9(d) and (e). The analyses are shown in Table 4. Inasmuch as the loss on ignition of a cement sample changes over a period of time, hence the apparent percentages of the various oxides, the values shown in this and succeeding tables have been calculated to the ignited (loss-free) basis. It should be recorded that all of these cements had been previously analyzed, four of them in these laboratories. The values shown are in good agreement with those previously obtained.

3. Determination of Aluminum Oxide in Portland Cements by the Proposed Methods

Table 5 shows the aluminum oxide content of the five cements determined by the proposed procedure with its three different methods (described above) for determining the aluminum oxide content of the oxyquinolate precipitate. The differences from the "by difference" values are also presented. Since each of the methods of estimating the percentage of the aluminum oxide, when used subsequent to the steps leading to the precipitation of aluminum oxyquinolate, constitutes an independent procedure, these methods are referred to in the discussion that follows as separate methods identified as I, II, and III.

TABLE 1—*ASTM procedures for portland cement.*

Oxide	ASTM Designation
"R_2O_3"	C 114-47, Section 9(a), (b), and (c).
Fe_2O_3	C 114-47, Section 11(a).
P_2O_5	C 114-47, Section 48(a), (b), and (c).
TiO_2	C 114-48T, Section 17, 18, and 19.

TABLE 2—Analysis of a synthetic solution (All values are grams of oxides in 25 mL of the solution).

Parameter	"R_2O_3"	Fe_2O_3	P_2O_5	TiO_2	By Difference	Al_2O_3 Direct Methods		
						I	II	III
Added	0.0550	0.0200	0.0030	0.0020	0.0300	0.0300	0.0300	0.0300
Found Test Number								
1	0.0553	0.0200	0.0031	0.0020	0.0302	0.0300	0.0295	0.0298
2	0.0553	0.0201	0.0031	0.0020	0.0301	0.0297	0.0295	0.0301
3	0.0547	0.0200	0.0031	0.0020	0.0296	0.0295	0.0295	0.0297
4	0.0548	0.0200	0.0031	0.0020	0.0297	...	...	...
5	0.0553	0.0201	0.0031	0.0020	0.0301	...	...	...
Average	0.0551	0.0200	0.0031	0.0020	0.0299	0.0297	0.0295	0.0299

METHODS USED:

R_2O_3—ASTM Designation C 114-47, Section 9(a), (b), and (c).
Fe_2O_3—ASTM Designation C 114-47, Section 11(a).
P_2O_5—ASTM Designation C 114-47, Section 48(a), (b), and (c).
TiO_2—ASTM Designation C 114-48T, Section 17, 18, and 19.
Al_2O_3, by difference—ASTM Designation C 114-47, Section 12.
Al_2O_3, direct methods:
I. Weighed as aluminum oxyquinolate; Al_2O_3 calculated.
II. Aluminum oxyquinolate titrated; Al_2O_3 calculated.
III. Aluminum oxyquinolate dissolved; aluminum precipitated as $Al(OH)_3$; ignited and weighed as Al_2O_3.

Methods I and II were developed with the thought that they could be used as rapid or alternate methods, accordingly, the experimental results shown in Table 5 were not corrected for blank determinations. Later work by other analysts showed that when corrected for blanks, Methods I and II were superior to Method III. It will be observed that in general, in spite of the lack of corrected values, good results were obtained by both Methods I and II. In the early stages of the work some quite high values were obtained. For example, Cement 18169 gave values of 7.14 and 6.97% Al_2O_3 with differences of 0.62 and 0.45%, respectively, from the gravimetric value of 6.52%. As the experience of the analyst increased, better results were obtained. Such high results (all uncorrected for blanks) would be omitted from the calculation of averages, were it not that results, obtained by other analysts inexperienced in these methods, showed similar high results when not corrected for blanks. It will be noted that the average values obtained by Methods I and II, except the average including the high values noted above and one other, are all in good agreement with the "by difference" values. Most of the individual values are within the ASTM C 114 limits of 0.20% between duplicate gravimetric determinations.

Most of the development work in this study was done with Method III using three of the five cements. Hence a considerable number of determinations are recorded for them. Since it was thought at first that this was the most likely procedure for use as a "standard" or "referee" method, blank tests were made and the amounts of silica present as a contaminant were determined. For comparison purposes both corrected and uncorrected values are shown in Table 5. It will be seen that most of the values obtained by Method III, both uncorrected and corrected, are within the ASTM limit of 0.20%. Considering the average values, all except the uncorrected average value for Cement 13895 are within the limit, and only one other varies by more than ±0.08% from the "by difference" values. Although the results

TABLE 3—*Reason cements were selected.*

Sample	Reason for Use in Tests
17596A	our laboratory reference sample; also identified as Check Sample 1
18169	high Al_2O_3 content
LTS 23	high Fe_2O_3 content
13895	high P_2O_5 content
18277	high TiO_2 content

in this table show that the corrected values are not necessarily more accurate, tests by other analysts, to be discussed later, indicated that the correction of the results obtained by Method III increased the accuracy.

4. Completeness of Separation of Aluminum from Iron, Phosphorus, and Titanium

A study was made to ascertain the amount of aluminum retained in the sodium hydroxide precipitate. Confirming the findings of Bright and Fowler [11], the loss of aluminum in the precipitate was found to be insignificant. To estimate the aluminum, the precipitate was ignited in a platinum crucible until the paper was completely consumed, the residue was fused with a small amount of potassium bisulfate, the melt was dissolved in water, the iron and titanium were separated by precipitation with sodium hydroxide, and the filtrate was acidified and, if necessary, concentrated to a smaller volume. Aluminum was determined on the filtrate by the colorimetric aurin tricarboxylic (aluminon) method [18]. Although all five cements were tested in triplicate, none of the values for retained aluminum oxide was in excess of 0.1 mg.

Tests of the ignited aluminum oxide obtained by decomposing the aluminum oxyquinolate precipitate with acids, precipitating the aluminum as hydroxide and igniting to the oxide, showed either the complete absence of iron, phosphorus, and titanium, or their presence

TABLE 4—*Analysis of portland cements by ASTM methods.*

Sample	R_2O_3	Fe_2O_3	P_2O_5	TiO_2	SiO_2	"By Difference" Al_2O_3
17596A	9.16	2.81	0.08	0.30	0.12	5.85
18169	9.83	2.86	0.03	0.28	0.14	6.52
LTS 23	10.50	5.51	0.28	0.36	0.06	4.29
13895	8.49	2.69	0.89	0.37	0.10	4.44
18277	10.45	4.89	0.10	0.77	0.18	4.51

METHODS USED:
    R_2O_3—ASTM Designation C 114-47, Section 9(a), (b), and (c).
    Fe_2O_3—ASTM Designation C 114-47, Section 11(a).
    P_2O_5—ASTM Designation C 114-47, Section 48(a), (b), and (c).
    TiO_2—ASTM Designation C 114-48T, Method C.
    SiO_2—ASTM Designation C 114-47, Section 9(d) and (e).
    Al_2O_3—ASTM Designation C 114-47, Section 12.

All values are shown as percentages of the ignited samples. All values were corrected for blanks and are the averages of 3 or more closely agreeing determinations.

TABLE 5—Analysis of portland cement for aluminum oxide by direct methods by analyst A.

	Sample Numbers																			
	17596A				18169				LTS 23				13895				18277			
	5.85				6.52				4.29				4.44				4.51			
	Al$_2$O$_3$ Percentages by ASTM "By Difference" Methods (Table 4)																			
Test	Not Corr.	Diff.	Corr.	Diff.	Not Corr.	Diff.	Corr.	Diff.	Not Corr.	Diff.	Corr.	Diff.	Not Corr.	Diff.	Corr.	Diff.	Not Corr.	Diff.	Corr.	Diff.
1	(6.35)	...	...	...	7.14	+.62	...	...												
2	5.93	+.08	...	...	6.97	+.45	...	...												
3	5.93	+.08	...	...	6.61	+.09	...	...												
4	5.92	+.07	...	...	6.66	+.14	...	...												
5	6.04	+.19	...	...	6.67	+.15	...	...												
6	5.93	+.08	...	...	6.62	+.10	...	...												
7					6.57	+.05	...	...												
8					6.60	+.08	...	...												
Ave	5.95	+.10			6.73	+.21														

	Al$_2$O$_3$ Percentages by Direct Methods																			
								METHOD I												
1									4.22	−.07	...	...	4.79	+.35			(5.15)	...	...	...
2									4.40	+.11	...	...	4.67	+.23			(5.11)	...	...	...
3									4.31	+.02	...	...	4.82	+.38			4.50	−.01	...	...
4									4.21	−.08	...	...	4.55	+.11			4.59	+.08	...	...
5									4.36	+.07	...	...	4.64	+.20			4.72	+.21	...	...
6									4.31	+.02	...	...	4.63	+.19			4.69	+.18	...	...
7									4.41	+.12	...	...	4.49	+.05			(5.97)	...	...	...
8									4.35	+.06	...	...	4.45	+.01			4.40	−.11	...	...
9													4.61	+.17			4.50	−.01	...	...
10													4.46	+.02						
Ave									4.33	+.04			4.61	+.17			4.57	+.06		
								METHOD II												
1	5.84	−.01	...	...	6.40	−.12	...	...	4.28	−.01	...	...	4.60	+.16			4.45	−.06	...	...
2	5.76	−.09	...	...	6.28	−.24	...	...	4.20	−.09	...	...	4.43	−.01			4.35	−.16	...	...
3	5.74	−.11	...	...	6.36	−.14	...	...	4.16	−.13	...	...	4.32	−.12			4.22	−.29	...	...
4	5.70	−.15	...	...	6.35	−.17	...	...	4.33	+.04	...	...	4.44	0			4.44	−.07	...	...
5	5.85	...	...	...	6.55	+.03	...	...	4.25	−.04	...	...	4.37	−.07			4.65	+.14	...	...
6	5.87	+.02	...	...	6.44	−.08	...	...	4.28	−.01	...	...	4.54	+.10			4.44	−.07	...	...

	Method I Not Corr.	Diff.	Corr.	Diff.	Method II Not Corr.	Diff.	Corr.	Diff.	Method III Not Corr.	Diff.	Corr.	Diff.	Not Corr.	Diff.	Corr.	Diff.	Not Corr.	Diff.	Corr.	Diff.	Not Corr.	Diff.	Corr.	Diff.
7	5.79	−.06			6.60	+.08							4.45	−.01							4.43	−.08		
8					6.60	+.08							4.48	+.04										
Ave	5.79				6.45	−.07			4.25	−.04			4.45	+.01							4.43			
1	6.04	+.19	5.92	+.07	6.71	+.19	6.56	+.04	(3.72)	…	(3.28)	…	4.53	+.24	4.35	+.06	4.75	+.31	4.75	+.02	4.37	−.14	4.21	−.30
2	5.90	+.05	5.82	−.03	6.75	+.23	6.61	+.09	4.53	+.24	4.35	+.06	4.07	−.22	3.99	−.30	4.75	+.31	4.48	+.04	4.31	−.20	4.23	−.28
3	6.02	+.17	5.92	+.07	6.75	+.23	6.63	+.11	4.07	−.22	3.99	−.30	4.43	+.14	4.39	+.10	4.63	+.19	4.59	+.15	(4.09)	…	(4.03)	…
4	5.90	+.05	5.82	−.03	6.75	+.23	6.61	+.09	4.43	+.14	4.39	+.10	4.39	+.10	4.35	+.06	(3.31)	…	(3.93)	…	4.39	−.12	4.33	−.18
5	5.92	+.07	5.84	−.01	6.79	+.27	6.67	+.15	4.39	+.10	4.35	+.06	4.01	−.28	3.95	−.34	(4.09)	…			4.19	−.32	4.13	−.38
6	6.04	+.19	5.96	+.11	6.79	+.27	6.63	+.11	4.01	−.28	3.95	−.34									4.29	−.34	4.19	−.32
6	6.00	+.15	5.90	+.10	6.85	+.33															4.27	−.24		
7	6.02	+.17	5.88	+.03	6.83	+.31															4.43	−.08		
8	5.98	+.13	5.86	+.01	6.61	+.09															4.33	−.18		
9	5.92	+.07	5.80	−.05	6.65	+.13															4.45	−.06		
10	5.92	+.07	5.80	−.05	6.40	−.12															4.57	+.06		
11	5.97	+.12	5.64	−.19	6.50	−.02															4.57	+.06		
12	(4.34)	…	(4.01)	…	6.30	−.22															(3.85)		(3.39)	
13	5.74	−.11	5.56	−.29	6.28	−.24	(5.79)	…													(4.06)			
14	6.03	+.18	5.84	−.01	6.50	−.02	6.44	−.08													4.75	+.24	4.71	+.20
15	5.74	−.11	5.54	−.31	6.67	+.12	6.61	+.09													4.71	+.20		
16					(6.12)		(6.08)														4.62	+.11	4.52	+.01
17																					4.44	−.06	4.33	−.18
Ave	5.94	+.09	5.80	−.05	6.63	+.09	6.60	+.08	4.29	0	4.21	−.08	4.71	+.27	4.51	+.07								

METHODS USED:

I. The aluminum oxyquinolate precipitate is dried to constant weight at 120 to 140°C., cooled and weighed.
II. The precipitate is dissolved in HCl and titrated with a standard solution of $KBrO_3$-KBr.
III. The precipitate is decomposed with acids, the aluminum precipitated as $Al(OH)_3$, ignited to Al_2O_3, and corrected for silica.

Values under headings "Not Corr." were not corrected for blanks or SiO_4.
Values under headings "Corr." were corrected for blanks by Methods I and II and for both blanks and SiO_2 by Method III.
Values under headings "Diff." are the differences between the direct method values and the "by difference" values.
All values were calculated to the "ignited" basis.
Values in parentheses ( ) were not included in the averages.

TABLE 6—*Impurities in the Al_2O_3 obtained by decomposing the aluminum oxyquinolate precipitate, precipitating the aluminum as $Al(OH)_3$, and igniting to Al_2O_o (Method III).*

Sample	Impurities in Al_2O_3				
	SiO_2	Fe_2O_3	P_2O_5	TiO_2	Total
17596A	0.06	faint	none	none	0.06
	0.06	trace	none	none	0.06
	0.06	trace	none	none	0.06
ave	0.06	trace	none	none	0.06
18169	0.04	faint	none	none	0.04
	0.04	trace	none	none	0.04
	0.04	trace	none	none	0.04
ave	0.04	trace	none	none	0.04
LTS 23	0.06	faint	none	none	0.06
	0.06	trace	none	none	0.06
	0.06	trace	none	none	0.06
ave	0.06	trace	none	none	0.06
13895	0.14	faint	none	none	0.14
	0.12	trace	none	none	0.12
	0.08	trace	none	none	0.08
ave	0.11	trace	none	none	0.11
18277	0.14	faint	none	less	0.14
	0.14	trace	none	than	0.14
	0.14	trace	none	0.01%	0.14
ave	0.14	trace	none	"	0.14

METHODS USED:

SiO_2—ASTM Designation C 114-47, Section 9(d) and (e).
Fe_2O_3—Colorimetric method.
P_2O_5—ASTM Designation C 114-47, Section 48(b) and (c).
TiO_2—ASTM Designation C 114-48T, Method C.

All values are corrected for blanks.
All values are percentages of the ignited samples.

in such minute quantities as to be wholly without effect on the determination. As shown in Table 6, of the impurities that may be present in the ignited aluminum oxide, only silicon dioxide was present in significant quantity. The fact that the sodium hydroxide method effected a good separation of iron and titanium from aluminum is evidenced by the presence of only a faint trace of iron and the complete absence of titanium in all except the sample that contains 0.77% TiO_2. In this instance, the amount of TiO_2 found in the ignited alumina is less than 0.01%. Phosphorus is absent indicating the completeness of separation of aluminum from phosphorus by precipitation with 8-hydroxyquinoline.

For the estimation of these impurities, the ignited alumina was fused with a small amount of potassium bisulfate and the melt dissolved in approximately 50 mL of distilled water. For the P_2O_5 determination, ASTM C 114-47, Section 48(b) and (c) was used. Iron was determined colorimetrically and TiO_2 according to ASTM C 114-48T, Method C.

5. Comparison of Results Obtained by the Proposed and Other Direct Methods

Chandler's [2] and Kampf's [3] procedures for the direct determination of aluminum were referred to earlier as being unsatisfactory. This conclusion is based on data presented in Table 7. The comparative tests were made on three of the five cements used in this study. Aluminum oxide values obtained "by difference" and by our proposed procedures are also

TABLE 7—Comparison of aluminum oxide values obtained by the proposed and other direct methods.

Method	Test	Sample		
		17596A	13895	18277
Grav. "by difference"	ave	5.85	4.44	4.51
Proposed No. I (Table 2)	ave (not corr.)	5.95	4.61	4.57
Proposed No. II (Table 2)	ave (not. corr.)	5.79	4.45	4.43
Proposed No. III (Table 2)	ave (corr.)	5.80	4.51	4.33
Chandler's gravimetric	1	5.18	4.02	4.04
	2	5.17	4.18	4.12
	3	5.27	4.16	4.10
	4	...	3.99	...
	ave	5.21	4.09	4.09
Chandler's volumetric	1	5.08	4.04	4.10
	2	5.12	4.07	4.10
	3	5.12	4.10	...
	ave	5.11	4.07	4.10
Kampf's volumetric	1	7.02	6.04	7.05
	2	7.02	6.04	7.10
	3	6.93		
	4	6.96		
	ave	6.98	6.04	7.08

METHODS USED:
    Proposed Methods I, II, and III.
    Chandler's gravimetric method (not corrected for blanks).
    Chandler's volumetric method (not corrected for blanks).
    Kampf's volumetric method (not corrected for blanks).
All values are percentages of Al_2O_3 calculated to the ignited basis.

shown in the table. It will be observed that Chandler's methods yielded low results. This confirms tests made some years ago by another analyst in these laboratories. The low results may be due to adsorption of a part of the aluminum by the bulky cupferron precipitate. Results by Kampf's method were much too high. The reasons for this are not entirely clear, although possible explanations are (1) the uncertainty of the end point due to gradual oxidation of the indicator and (2) the presence of titanium, an interfering element, in the precipitate. It is also possible that some necessary details of the procedure that Kampf used in obtaining his excellent results were omitted from the published procedure.

6. Comparison of Results Obtained by Different Analysts Using Proposed Methods

To ascertain the reproducibility of results obtained by the direct method, the five cements were analyzed by three other analysts using the same three methods for determining the aluminum oxide content of the oxyquinolate precipitate. The comparative results, including average values obtained by the four analysts, are shown in Table 8. The individual values obtained by Analyst A are the ones that appeared in Table 5 and are not repeated in Table 8. All values, however, were included in studying the data.

The table shows that in general the analysts, within tolerance limits, were able to reproduce their own results as shown by the individual values. Reproducibility was better with the corrected than with the uncorrected values. The accuracy of their corrected average results was good, as shown by the differences from the "by difference" values. The uncorrected results were inaccurate in a number of instances.

TABLE 8—*Analysis of portland cements for aluminum oxide in direct methods by four analysts.*

		\multicolumn{4}{c}{Sample Numbers}															
		\multicolumn{4}{c}{17596A}	\multicolumn{4}{c}{18169}	\multicolumn{4}{c}{LTS 23}	\multicolumn{4}{c}{13895}	\multicolumn{4}{c}{18277}
Analyst	Test	Not Corr.	Diff.	Corr.	Diff.	Not Corr.	Diff.	Corr.	Diff.	Not Corr.	Diff.	Corr.	Diff.	Not Corr.	Diff.	Corr.	Diff.	Not Corr.	Diff.	Corr.	Diff.
		\multicolumn{20}{c}{Al_2O_3 Percentages by ASTM "By Difference" Methods (Table 4)}
		\multicolumn{4}{c}{5.85}	\multicolumn{4}{c}{6.52}	\multicolumn{4}{c}{4.29}	\multicolumn{4}{c}{4.44}	\multicolumn{4}{c}{4.51}
		\multicolumn{20}{c}{Al_2O_3 Percentages by Direct Methods}
		\multicolumn{20}{c}{METHOD I}
A	ave	5.95	+0.10	...	...	...	+0.21	...	...	4.33	+0.04	...	...	4.61	+0.17	...	...	4.57	+0.06	...	...
B	1	5.93	+0.08	5.80	−0.05	6.73	+0.21	6.47	−0.05	4.41	+0.12	4.27	−0.02	4.70	+0.26	4.56	+0.12	4.55	+0.04	4.42	−0.09
	2	5.86	+0.01	5.74	−0.11	6.56	+0.04	6.43	−0.09	4.47	+0.18	4.25	−0.04	4.55	+0.11	4.52	+0.08	4.62	+0.11	4.48	−0.03
	3	6.00	+0.15	5.86	+0.01	6.65	+0.13	6.51	−0.01	4.48	+0.19	4.35	+0.06	4.80	+0.36	4.67	+0.23	4.67	+0.16	4.54	+0.03
	ave	5.93	+0.08	5.80	−0.05	6.61	+0.09	6.47	−0.05	4.46	+0.17	4.29	0	4.68	+0.24	4.58	+0.14	4.61	+0.10	4.48	−0.03
C	1	6.72	+0.37	5.90	+0.05	...	...	(6.24)	...	4.70	+0.41	4.40	+0.11	4.98	+0.54	4.67	+0.23	4.72	+0.21	4.41	−0.10
	2	6.28	+0.43	5.99	+0.14	6.91	+0.39	6.63	+0.11	4.80	+0.51	4.52	+0.23	5.02	+0.58	4.73	+0.29	4.80	+0.29	4.51	0
	3	6.11	+0.26	5.86	+0.01	6.79	+0.27	6.54	+0.07	4.56	+0.27	4.32	+0.03	4.88	+0.44	4.63	+0.19				
	4	6.16	+0.31	5.81	−0.04	7.02	+0.50	6.77	+0.25					4.96	+0.52	4.71	+0.27				
	5					6.87	+0.35	6.62	+0.10												
	ave	6.19	+0.34	5.52	+0.07	6.90	+0.38	6.64	+0.12	4.69	+0.40	4.41	+0.12	4.96	+0.52	4.69	+0.27	4.76	+0.25	4.46	−0.05
	no of detns.	12		7		15		7		14		6		17		7		11		5	
	ave of all	6.03	+0.18	5.87	+0.07	6.75	+0.23	6.57	+0.05	4.44	+0.15	4.35	+0.06	4.70	+0.26	4.64	+0.20	4.61	+0.10	4.47	−0.04
	maximum value	6.28	+0.43	5.99	+0.14	7.14	+0.62	6.77	+0.25	4.80	+0.51	4.52	+0.23	5.02	+0.58	4.73	+0.29	4.80	+0.29	4.54	+0.03
	minimum value	5.86	+0.01	5.74	−0.11	6.56	+0.04	6.43	−0.09	4.21	−0.08	4.25	−0.04	4.45	+0.01	4.52	+0.08	4.40	−0.11	4.41	−0.10
	standard error	0.27		0.07		0.29		0.12		0.21		0.11		0.32		0.21		0.15		0.06	
		\multicolumn{20}{c}{METHOD II}
A	ave	5.79	−0.06	...	...	...	−0.07	...	...	4.25	−0.04	...	...	4.45	+0.01	...	...	4.43	−0.08	...	...
B	1	5.78	−0.07	5.61	−0.24	6.45	−0.07	6.26	−0.26	4.35	+0.06	4.17	−0.12	4.60	+0.16	4.42	−0.02	4.42	−0.09	4.25	−0.26
	2	5.75	−0.10	5.57	−0.28	6.44	−0.08	6.26	−0.26	4.32	+0.03	4.15	−0.14	4.58	+0.14	4.40	−0.04	4.58	+0.07	4.38	−0.13
	3	5.88	+0.03	5.72	−0.13	6.49	−0.03	6.33	−0.19	4.40	+0.11	4.23	−0.06	4.71	+0.27	4.53	+0.09	4.58	+0.07	4.42	−0.09
	ave	5.80	−0.05	5.63	−0.22	6.46	−0.06	6.28	−0.24	4.36	+0.07	4.18	−0.11	4.63	+0.19	4.45	+0.01	4.53	+0.02	4.35	−0.16
C	1	5.97	+0.12	5.74	−0.11	6.63	+0.09	6.40	−0.12	...	...	(4.10)	...	4.88	+0.44	4.59	+0.15	4.87	+0.36	4.49	−0.02
	2	(6.38)	...	(6.01)	...	6.77	+0.25	6.40	−0.12	4.58	+0.29	4.36	+0.07	4.79	+0.35	4.57	+0.13	4.74	+0.23	4.45	−0.06
	3	(5.70)	...	(5.40)	...	...	...	(6.03)	...	4.66	+0.37	4.40	+0.11	4.86	+0.42	4.59	+0.15	4.64	+0.13	4.41	−0.10

	4	6.01	+0.16	5.78	−0.07	6.69	+0.17	6.43	−0.09	4.62	+0.33	4.36	+0.07	4.84	+0.40	4.58	+0.14	4.75	+0.24	4.45	−0.06
	5		+0.22	5.80	−0.05																
	ave	6.07	+0.17	5.77	−0.12	6.67	+0.15	6.42	−0.10	4.62	+0.33	4.37	+0.08							6	
	no. of detns.	6.02																			
		12		6		14		6		12		6		14		6		12			
	ave of all	5.85	0	5.70	−0.15	6.50	−0.02	6.35	−0.17	4.37	+0.08	4.28	−0.01	4.57	+0.13	4.52	+0.08	4.54	+0.03	4.40	−0.11
	maximum value	6.07	+0.22	5.80	−0.05	6.77	+0.25	6.43	−0.09	4.66	+0.37	4.40	+0.11	4.88	+0.44	4.59	+0.15	4.87	+0.36	4.49	−0.02
	minimum value	5.70	−0.15	5.57	−0.28	6.28	−0.24	6.26	−0.26	4.16	−0.13	4.15	−0.14	4.32	−0.12	4.40	−0.04	4.22	−0.29	4.25	−0.26
	standard error		0.11		0.17		0.13		0.19		0.18		0.10		0.22		0.11		0.17		0.13

METHOD III

A	ave	5.94	+0.09	5.80	−0.05	6.63	+0.09	6.60	+0.08	4.29	0	4.21	−0.08	4.71	+0.27	4.51	+0.07	4.44	−0.07	4.33	−0.18
B	1	5.81	−0.04	5.67	−0.18	6.47	−0.05	6.29	−0.23	4.45	+0.16	4.25	−0.04	4.55	+0.11	4.41	−0.03	4.47	−0.04	4.33	−0.18
	2	6.10	+0.25	5.81	−0.04			6.23	−0.29	4.39	+0.10	4.19	−0.10	4.55	+0.11	4.39	−0.05	4.47	−0.04	4.35	−0.16
	3	6.12	+0.27	5.93	+0.08	(7.19)		6.37	−0.15	4.65	+0.36	4.46	+0.17	4.66	+0.22	4.46	+0.02	4.51	0	4.37	−0.14
	ave	6.01	+0.16	5.80	−0.05	6.53	+0.01	6.30	−0.22	4.50	+0.21	4.30	+0.01	4.59	+0.15	4.42	−0.02	4.48	−0.03	4.35	−0.16
C	1	6.30	+0.45	5.95	+0.10	6.50	−0.02	6.68	+0.16	4.78	+0.49	4.38	+0.09	5.00	+0.56	4.65	+0.21	5.14	+0.63	4.74	+0.23
	2	6.28	+0.43	5.89	+0.05	7.07	+0.55	6.60	+0.08	4.64	+0.35	4.28	−0.01	5.00	+0.56	4.61	+0.17	4.91	+0.40	4.53	+0.02
	3	6.08	+0.23	5.71	−0.14	6.99	+0.47	6.54	+0.02	4.72	+0.43	4.36	+0.07	5.23	+0.79	4.86	+0.42	4.93	+0.42	4.53	+0.02
	4	6.02	+0.17	5.67	−0.18	6.91	+0.39	6.62	+0.10	4.52	+0.23	4.15	−0.14	5.15	+0.71	4.80	+0.36				
	ave	6.17	+0.32	5.81	−0.04	6.97	+0.45	6.61	+0.09	4.67	+0.38	4.29	0	5.10	+0.66	4.73	+0.29	4.99	+0.48	4.60	+0.09
D	1	6.13	+0.28	5.82	−0.03	6.99	+0.47	(6.05)		4.65	+0.36	4.39	+0.10	4.79	+0.35	4.40	−0.04	4.98	+0.47	4.62	+0.11
	2	6.11	+.26	5.84	−0.01			6.61	+0.09	4.57	+0.28	4.43	+0.14	4.83	+0.39	4.61	+0.17	4.61	+0.10	4.22	−0.29
	3	6.01	+0.16	5.70	−0.15	6.83	+0.31	6.26	−0.26			(4.01)		5.10	+0.66	4.69	+0.25	4.73	+0.22	4.36	−0.15
						6.63	+0.11	6.40	−0.12					5.10	+0.66	4.		(4.40)		(4.02)	
						6.75	+0.23									+0.37					
	ave	6.08	+0.23	5.79	−0.06	6.74	+0.22	6.42	−0.10	4.61	+0.32	4.41	+0.12	4.96	+0.52	4.63	+0.19	4.77	+0.26	4.40	−0.11
	no. of detns.	25		25		25		18		14		14		14		14		23		16	
	ave of all	6.00	+0.15	5.80	−0.05	6.69	+0.17	6.52	0	4.49	+0.19	4.28	−0.01	4.87	+0.43	4.59	+0.15	4.56	+0.05	4.40	−0.11
	maximum value	6.30	+0.45	5.96	+0.11	7.07	+0.55	6.68	+0.16	4.78	+0.49	4.43	+0.14	5.23	+0.79	4.86	+0.42	5.14	+0.63	4.74	+0.23
	minimum value	5.74	−0.11	5.54	−0.31	6.28	−0.24	6.23	−0.29	4.01	−0.28	3.95	−0.34	4.55	+0.11	4.39	−0.05	4.17	−0.34	4.13	−0.38
	standard error		0.20		0.12		0.26		0.14		0.29		0.15		0.48		0.21		0.26		0.21

METHODS USED:

I. The aluminum oxyquinolate precipitate is dried to constant weight at 120 to 140°C., cooled, and weighed.
II. The precipitate is dissolved in HCl and titrated with a standard solution of $KBrO_3$-KBr.
III. The precipitate is decomposed with acids, the aluminum precipitated as $Al(OH)_3$, ignited to Al_2O_3, and corrected for silica.

Average analyses by Analyst A (Table 5) are included for reference. His individual values, although not shown in this table, were used in computing averages of all, differences and maximum and minimum values.

Values under headings "Not Corr." were not corrected for blanks or SiO_2.
Values under headings "Corr." were corrected for blanks by Methods I and II and for both blanks and SiO_2 by Method III.
Values under headings "Diff." are the difference between the direct method values and "by difference" values.
All values were calculated to the "ignited" basis.
Values in parentheses ( ) were not included in the averages.
NOTE: Standard error = $\sqrt{\Sigma d^2/n}$, where d = difference from "by difference" values, and n = number of determinations.

7. Comparison of the Proposed Methods

It appeared from a study of Table 8 that the best estimate of the worth of the methods could be made by considering the results obtained by all analysts rather than only those of Analyst A who developed them. In the following discussion all the results for a given method and sample are considered as though obtained by one person.

One estimate of the accuracy of the methods may be obtained from the differences from the "by difference" aluminum oxide values. The number of determinations for each cement and the average differences are shown in Table 9. It was mentioned earlier that the "by

TABLE 9—*Summary of mathematical data.*

Cement	Method I		Method II		Method III	
	Not Corr.	Corr.	Not Corr.	Corr.	Not Corr.	Corr.
	NUMBER OF DETERMINATIONS					
1759A	12	7	12	6	25	25
18169	15	7	14	6	25	18
LTS 23	14	6	12	6	14	14
13895	17	7	14	6	14	14
18277	11	5	12	6	23	16
Total	69	32	64	30	101	87
	DIFFERENCES—AVERAGES FOR FOUR ANALYSTS, %					
17596A	+0.18	+0.02	0	−0.15	+0.15	−0.05
18169	+0.23	+0.05	+0.02	−0.17	+0.17	0
LTS 23	+0.15	+0.06	+0.08	−0.01	+0.19	−0.01
13895	+0.26	+0.20	+0.13	+0.08	+0.43	+0.15
18277	+0.10	−0.04	+0.03	−0.11	+0.05	−0.13
Weighted average	+0.19	+0.06	+0.04	−0.07	+0.18	−0.02
	STANDARD[1] AND PROBABLE[2] ERRORS FOR FOUR ANALYSTS WITH RESPECT TO "BY DIFFERENCE" VALUES, %					
17596A	0.22	0.07	0.11	0.17	0.20	0.12
18169	0.29	0.12	0.13	0.19	0.26	0.14
LTS 23	0.21	0.11	0.18	0.10	0.29	0.15
13895	0.32	0.21	0.22	0.11	0.48	0.21
18277	0.15	0.06	0.17	0.13	0.26	0.21
S.E. for all	0.26	0.13	0.17	0.16	0.29	0.17
P.E. for all	0.17	0.08	0.11	0.10	0.20	0.11
	PROBABLE ERRORS[3] OF FOUR ANALYSTS WITH RESPECT TO AVERAGE DIRECT VALUES, %					
17596A	0.09	0.05	0.08	0.06	0.09	0.08
18169	0.13	0.08	0.09	0.06	0.14	0.10
LTS 23	0.12	0.07	0.11	0.06	0.15	0.11
13895	0.12	0.05	0.11	0.06	0.16	0.11
18277	0.08	0.04	0.12	0.06	0.17	0.13
P.E. for all	0.11	0.06	0.10	0.06	0.14	0.10

NOTE: (1) Standard error = $\sqrt{\Sigma d^2/n}$, where d = difference from "by difference" values, and n = number of determinations. (2) Probable error with respect to "by difference" values = 0.6745 times standard error. (3) Probable error with respect to average

$$= 0.6745 \sqrt{\frac{\Sigma x^2 - (n_1 m_1^2 \ldots n_5 m_5^2)}{(n_1 \ldots n_5) - 5}}$$

where x = individual values,
n_1 etc. = number of determinations for each cement, and
m_1 etc. = average of n_1 etc., determinations.

difference" values obtained by ASTM methods were subject to cumulative errors. However, since such values at present are considered as being sufficiently accurate for acceptance or referee tests, the "by difference" values are assumed to be correct in the following evaluations of the proposed methods. The weighted average differences appearing in Table 9 show that (1) the corrected results for Methods I and II are better than the corresponding uncorrected results and (2) that both the uncorrected and corrected average differences for Method II are low, therefore very good.

A better estimate of relative accuracy may be obtained by calculating standard errors with respect to the "by difference" values. Such values, also shown in Table 9, indicate that the relative accuracy of all methods, corrected, is good. It will be seen that for all methods the errors for the corrected values, with two exceptions, are less than for the corresponding uncorrected values. Considering the standard errors for all cements, Method II gave the lowest errors for uncorrected values.

The probable error values with respect to average direct method values, also shown in Table 9, show the precision of all the methods to be good. Results by Methods I and II for all determinations, corrected, show the lowest probable errors, namely 0.06% in each case. The highest probable error for all determinations was 0.14% by Method III, uncorrected.

In considering the worth of the methods, the time requirements should be considered also. As previously stated, when circumstances are such as to require that Al_2O_3 values as commonly calculated be corrected for P_2O_5 and TiO_2, it is necessary, according to the current ASTM methods, to determine the amounts of these components separately and subtract them from the gross Al_2O_3 values. All laboratories analyzing cement regularly are set up to determine "R_2O_3" and Fe_2O_3. The time required can be assumed to be only a few hours, starting with the NH_4OH group precipitate for the "R_2O_3" and dry cement for the Fe_2O_3. It is difficult, however, to estimate the time requirements for the determination of TiO_2 since there are several ASTM tentative (no referee) methods differing greatly in length and complexity. The time required for the determination of P_2O_5 by the referee method is at least 24-h elapsed time. Considerable time may be saved if the use of the alternate method for P_2O_5 is permitted. While the author has not made a time study, it is his experience that 16 to 28 working hours would be required to make all the above determinations, the time varying with the number of samples and the availability of apparatus and solutions of reagents.

Assuming again that the ammonium hydroxide group has been separated, the approximate elapsed time required for each method is shown below. The time varies with the experience of the analyst and the number of samples being analyzed.

From a time standpoint, it will be seen that either Method I or II is preferable to Method III (Table 10). Even Method III is faster than the ASTM method. It was found that Method II was the most rapid for one or two samples provided all solutions have been prepared previously. For occasional tests, however, Method I is the most rapid. It is also the simplest for the analyst. On the other hand, Method II is the least afffected by possible impurities in the oxyquinolate precipitate. This was shown in the mathematical studies discussed above in which it was brought out that results by this procedure were good even when uncorrected for blanks.

TABLE 10—*Comparison of methods.*

Method	Time, Elapsed Hours
I	6 to 9, (includes some "waiting" time)
II	6 to 10, (does not include preparation and standardization of volumetric solutions)
III	12 to 18, (this includes considerable "waiting" time during which other work may be carried on)

Summary

1. A chemical method for the direct determination of aluminum oxide in portland cement has been developed.

2. By using the precipitate of the ammonium hydroxide group for the determination, the method can be incorporated as a part of the ASTM procedures for analysis of portland cement.

3. After the separation of aluminum as oxyquinolate, the aluminum oxide content of the precipitate may be determined in any one of these ways:

- Method I—The aluminum oxyquinolate precipitate is dried and weighed and the aluminum oxide content calculated.
- Method II—The aluminum oxide content of the precipitate is determined volumetrically.
- Method III—The precipitate is subjected to wet oxidation of the organic matter and aluminum is separated as the hydroxide and ignited and weighed as the oxide.

4. A synthetic solution containing Al_2O_3, Fe_2O_3, TiO_2, and P_2O_5 was analyzed by each method with good results.

5. Five commercial portland cements were analyzed by each method. Studies of various precipitates and filtrates showed that interference by other elements of the ammonium hydroxide group was eliminated.

6. Comparative tests showed the procedure to be superior to two other published methods.

7. Reproducibility of results by other analysts was good provided corrections were made for blank determinations.

8. A mathematical study of the data showed accuracy and precision of all methods was good when corrections were made for blank determinations, but that Methods I and II were more accurate and precise than Method III. Method II is the least subject to errors because of possible contamination of the oxyquinolate precipitate.

9. On a time basis, Method I is the fastest for an occasional sample, and Method II is the fastest as a routine procedure. In cases where it is required that the Al_2O_3 values be corrected for P_2O_5 and TiO_2, the direct methods are more rapid than the current ASTM procedures.

Acknowledgment

Mr. W. S. Lui, formerly Associate Research Chemist, Research and Development Division, Portland Cement Association, was responsible for a large share of the early work on the developments reported in this paper. Acknowledgment of his contributions is herewith extended.

APPENDIX

I. Preliminary Separations

Section 1

(a) Separate but do not weigh (unless desired) the silicon dioxide from a 0.500-g sample of cement according to ASTM Designation C 114-47, Section 33(a) and (b). Volatilize the silicon dioxide thus obtained and recover the residue according to Section 8(d).

(b) Precipitate the ammonium hydroxide group according to Section 9(a) and (b) and treat the precipitate according to 9(c). Discard the filtrate (unless a determination of CaO and MgO is desired).

II. Separation of Aluminum as the Oxyquinolate

Section 2

Reagents

(a) *Sodium Hydroxide Solution (10%)*. Dissolve 100 g of C. P. sodium hydroxide in distilled water and dilute to 1 L.

(b) *Sodium Hydroxide Wash Solution (5%)*. Dissolve 50 g of C. P. sodium hydroxide and 5 g of C. P. sodium sulfate in distilled water and silute to 1 L.

(c) *8-Hydroxyquinoline Solution (2.5%)*. Treat 12.5 g of 8-hydroxyquinoline with 25 mL of C. P. glacial acetic acid, and warm gently to effect solution. Pour the resulting solution into the 450 mL of distilled water at 60°C. Cool, filter if necessary, and dilute to 500 mL.

(d) *Ammonium Acetate Solution*. Dissolve 100 g of C. P. ammonium acetate in 100 mL of distilled water.

Procedure

(a) Place the precipitate of the ammonium hydroxide group into a platinum crucible of approximately 40-mL capacity. Dry and ignite the papers, first at a low heat until the carbon of the paper is complete consumed without inflaming, and finally at 1050 to 1100°C for 10 min. Add 5 g of fused C. P. $KHSO_4$ to the crucible, and heat below red heat until the residue is dissolved in the melt (Note 1). Cool, dissolve the fused mass in approximately 50 mL of distilled water and 5 mL of H_2SO_4 (1:1) in a 250-mL beaker. Nearly neutralize with the sodium hydroxide solution (10%) (Note 2), adjust the volume to about 100 mL, heat nearly to boiling, and pour slowly into 100 mL of a hot sodium hydroxide solution (10%) as the latter is constantly stirred. Rinse out the beaker several times with small portions of hot water and add to the solution. Boil for 2 to 3 min, let settle on a steam bath for about 15 min, filter through an 11-cm. No. 41 Whatman paper (or equivalent) into a 600-L beaker, and wash the paper and precipitate several times with small portions of hot sodium hydroxide wash solution.

NOTE 1—Start the heating slowly and with caution to prevent foaming and spattering.
NOTE 2—About 50 mL are required.

(b) The combined filtrate and washings should have a volume of about 275 mL. Add HCl (sp gr 1.18) until the solution is just acid to methyl red. Heat the solution, and add dilute ammonia (1:1) until one drop just changes the color of the solution to yellow. At once add dilute HCl (1:1), drop by drop, until the solution is again red and the precipitated $Al(OH)_3$ is just dissolved (Note 3). Cool the solution somewhat then add 5 mL of acetic acid (1:1) and 15 mL of the 8-hydroxyquinoline solution (Note 4). Finally add, slowly and with stirring, 20 mL of the ammonium acetate solution. Heat the solution to 60 to 70°C and digest at this temperature for 5 min to facilitate crystallization and coagulation of the precipitate. Allow the precipitate to settle for 15 min, while cooling to room temperature (Note 5).

NOTE 3—The presence of any $Al(OH)_3$ should be carefully checked at this point to be sure it is all in solution.

NOTE 4—One millilitre of the 8-hydroxyquinoline reagent suffices to precipitate 2.9 mg of alumina. An excess of the reagent does no harm; in any case enough should be used to color the solution yellow.

NOTE 5—The precipitate should be filtered within 1 h. Prolonged standing may cause high results.

(c) Determine the aluminum oxide content of the precipitate by either Section 4, 5, or 6.

III. Determination of Aluminum oxide

Section 4

The precipitate is filtered, washed, dried, and weighed as anhydrous aluminum oxyquinolate, $Al(C_9H_6ON)_3$.

Procedure

(a) Filter the precipitate, using moderate suction, through a weighed 30-mL fritted glass crucible of fine porosity. Wash the precipitate with warm NH_4OH (1:40) until the washings are colorless, dry for 1½ to 2 h at 120 to 140°C. Cool and weigh as anhydrous aluminum oxyquinolate.

(b) *Blank.* Make a blank determination following the same procedure and using the same amounts of reagents, and correct the results obtained in the analysis accordingly.

(c) *Calculation.* Calculate the percentage of Al_2O_3 to the nearest 0.01 as follows

$$Al_2O_3\% = W \times 22.198$$

where

W = weight of anhydrous $Al(C_9H_6ON)_3$, and
22.198 = molecular ratio of Al_2O_3 to $Al(C_9H_6ON)_3$ (0.11099) divided by the sample weight (0.5) and multiplied by 100.

Section 5

The precipitate is titrated, after solution in hydrochloric acid, with a standard solution of $KBrO_3$-KBr.

Reagents

(a) *Standard Sodium Thiosulfate Solution (0.35N).* Dissolve 88 g of C. P. sodium thiosulfate ($Na_2S_2O_3 \cdot 5H_2O$) in 300 mL of recently distilled water, add 0.1 g of C. P. sodium carbonate, and dilute to 1 L. Standardize this solution against 0.1 N $K_2Cr_2O_7$ as follows:

Prepare a 0.1 N solution of $K_2Cr_2O_7$ by dissolving 4.904 g of Bureau of Standards Sample 136 in water and diluting to 1 L. Measure accurately 50 mL of the solution into a 500-mL Erlenmeyer flask and dilute with water to about 250 mL. Add 25 mL of HCl (sp gr 1.18), shake, then add 15 mL of KI (25%). Shake again, and titrate with the thiosulfate solution until the solution is nearly colorless. At this point add 1 or 2 mL of starch solution and titrate to the disappearance of the blue color

$$\text{normality of } Na_2S_2O_3 \text{ solution} = 0.1 \times \frac{\text{mL of } K_2Cr_2O_7}{\text{mL of } Na_2S_2O_3}$$

(b) *Standard Potassium Bromate-Potassium Bromide Solution (0.35 N).* Dissolve 9.743 g of C. P. potassium bromate and approximately 34 g of C. P. potassium bromide in 400 mL of distilled water and dilute to 1 L. Obtain the ratio of the strength of this solution to that of the standard $Na_2S_2O_3$ solution as follows: To 250-mL of distilled water in an Erlenmeyer flask add exactly 25 mL of the bromate-bromide solution, shake, add 25 mL of HCl (sp gr 1.18), and immediately add 15 mL of a 25% solution of KI. Shake and titrate with the thiosulfate solution until the solution is nearly colorless. At this point add 2 mL of starch solution, and titrate to the disappearance of the blue color.

Calculate the Al_2O_3 value of the $KBrO_3$-KBr solution and the $KBrO_3$-KBr equivalent of the $Na_2S_2O_3$ solution as follows:

$$\text{normality of } KBrO_3\text{-KBr solution} = \text{normality of } Na_2S_2O_3 \text{ solution} \times \frac{\text{mL of } Na_2S_2O_3}{\text{mL of } KBrO_3\text{-KBr}}$$

$$\text{grams } Al_2O_3 \text{ per mL of } KBrO_3\text{-KBr solution} = \text{normality of } KBrO_3\text{-KBr} \times 0.0042485$$

where 0.0042485 is the Al_2O_3 equivalent of 1 mL of 0.1 N solution of $KBrO_3$-KBr

$$KBrO_3\text{-KBr equivalent of the } Na_2S_2O_3 \text{ solution} = \frac{\text{normality of } Na_2S_2O_3}{\text{normality of } KBrO_3\text{-KBr}}$$

(c) *25% Potassium Iodide Solution.* Dissolve 25 g of C. P. potassium iodide in 100 mL of distilled water.

(d) *Starch Solution.* To 500 mL of boiling water, add a cold suspension of 5 g of soluble starch in 25 mL of water. Cool, add a cool solution of 5 g of C. P. sodium hydroxide in 50 mL of water, then add 15 g of C. P. potassium iodide, and mix thoroughly.

Procedure

(a) Filter the solution containing the oxyquinolate precipitate by suction through a 30-mL fritted glass crucible of fine porosity. Wash the beaker and precipitate with about 60-mL diluted ammonium hydroxide (1:99) heated to about 50°C. Place the crucible on a clean filter flask and pour 25 mL of hot (75°C) diluted HCl (1:6) on the washed precipitate. Let the reaction proceed for a few minutes, and stir occasionally with a small glass rod before applying suction. As soon as the precipitate has dissolved apply suction and wash the crucible twice with 75-mL portions of hot diluted HCl (75°C) then with 50 mL of water. Dilute the acid solution to 250 mL, add 15 mL of HCl, and cool to room temperature (21 ± 4°C). The hydrochloric acid content of the solution during the subsequent bromination should not be less than 8%. Add 25 mL of the standard $KBrO_3$-KBr solution from a buret or pipet. Stir the solution and allow to stand for 30 s to insure complete bromination. Add 10 mL of 25% KI. Stir the resulting solution well, and the titrate very slowly with the standard $Na_2S_2O_3$ solution until the color of the iodine becomes faintly yellow. At this point add 2 mL of the starch solution, and titrate to the disappearance of the blue color.

(b) *Blank.* Make a blank determination following the same procedure and using the same amounts of reagents, and correct the results obtained in the analysis accordingly.

(c) *Calculation.* Calculate the percentage of Al_2O_3 to the nearest 0.01 as follows:

Correct the amount of $KBrO_3$-KBr by subtracting from it the $KBrO_3$-KBr equivalent of the $Na_2S_2O_3$ used

$$Al_2O_3\% = V \times F \times 200$$

where

V = number of millilitres of $KBrO_3$-KBr solution,
F = grams of Al_2O_3 per millilitre of $KBrO_3$-KBr solution, and
200 = 100 divided by the sample weight (0.5 g).

Section 6

The precipitate is decomposed with sulfuric and nitric acids, the aluminum precipitated as hydroxide, ignited to the oxide, weighed, and corrected for silica.

Procedure

(a) Filter through a Whatman 41, 11-cm (or equivalent) filter paper into a 600-mL beaker the solution containing the oxyquinolate precipitate, and wash the precipitate with several small portions of hot NH_4OH(1:40). Digest the paper and precipitate in an excess (about 30 mL), of sulfuric acid (1:1), evaporate to fumes of that acid, and while fuming add successive 5-mL portions of nitric acid (HNO_3) (sp gr 1.42) to destroy the bulk of the organic matter. Cool the solution slightly, add 10 mL of HNO_3 (sp gr 1.42), and continue heating until the solution is colorless. Cool, dilute with water to about 200 mL, and boil until the anhydrous aluminum sulfate is completely dissolved. Cool the solution slightly, add a few drops of methyl red indicator, then treat slowly with freshly filtered NH_4OH (1:1) until the color of the solution becomes yellow and add one drop in excess. Bring the solution to boiling and boil for 10 to 15 s. Allow the precipitate to settle and filter. Wash four times with hot NH_4Cl (20 g per L.). Place the paper and precipitate into a weighed platinum crucible. Dry and ignite the paper, first at low heat until the carbon of the paper is completely consumed without inflaming, and finally at 1050 to 1100°C until the weight remains constant. The difference between this weight and the weight of the empty crucible represents the weight of Al_2O_3 plus traces of SiO_2.

(b) Add 5 g of C. P. fused $KHSO_4$ to the crucible, and heat below red heat until the residue is dissolved in the melt. Cool, dissolve the fused mass in water containing 5 mL of H_2SO_4 (1:1), and evaporate the solution. Raise the temperature until copious fumes just begin to be evolved. Dissolve the mass in water, digest for 15 to 30 min at a temperature just below the boiling point, filter, and wash with hot water. Transfer the paper containing the residue to a platinum crucible, heat slowly until the paper is charred, and finally ignite to constant weight at 1050 to 1100°C. Treat the SiO_2 thus obtained in the crucible with a drop of water, about 5 mL of HF, and a drop of H_2SO_4, and evaporate cautiously to dryness. Finally, heat the crucible at 1050 to 1100°C for 10 min, cool, and weigh. The difference between this weight and the first weight represents the amount of SiO_2. Subtract this weight from the weight of Al_2O_3 obtained.

(c) *Blank.* Make a blank determination following the same procedure and using the same amounts of reagents and correct the results obtained in the analysis accordingly.

(d) *Calculation.* Calculate the percentage of Al_2O_3 to the nearest 0.01 as follows

$$Al_2O_3\% = \frac{\text{wt of } Al_2O_3}{\text{wt of sample}} \times 100$$

References

[1] Ford, C. L. and LeMar, L., *ASTM Bulletin*, No. 157, March 1949, p. 66.
[2] Chandler, W. R., *Rock Products*, Vol. 39, No. 8, 1936, p. 52.
[3] Kampf, L., *Ind. Eng. Chem., Anal. Ed.*, Vol. 13, 1941, p. 72.
[4] Lundell, Hoffman, and Bright, *Chemical Analysis of Iron and Steel*, John Wiley & Sons, New York, 4th Printing, 1946, p. 78.
[5] Willard, H. H. and Tang, N. K., *Ind. Eng. Chem., Anal. Ed.*, Vol. 9, 1937, p. 357.
[6] Boyle, A. J. and Musser, D. F., *Ind. Eng. Chem., Anal. Ed.*, Vol. 15, 1943, p. 621.
[7] Kassner, J. L. and Ozier, M. A., *Journal of the American Ceramics Society*, Vol. 33, No. 8, 1950, pp. 250–252.
[8] Welcher, F. J., *Organic Analytical Reagents*, Vol. IV, D. Van Nostrand Co., New York, 1948, p. 159.
[9] Smith, G. F. and Cagle, F. W., *Ind. Eng. Chem., Anal. Ed.*, Vol. 20, 1948, p. 574.
[10] Hillebrand, W. F. and Lundell, G. E. F., "Applied Inorganic Analysis," John Wiley & Sons, New York, 1929, p. 76.
[11] Bright, H. A. and Fowler, R. M., *Journal of Research NBS*, Vol. 10, 1933, p. 330.
[12] Kolthoff, I. M., Stenger, V. A., and Moskovitz, B., *Journal of the American Ceramic Society*, Vol. 56, 1934, p. 812.
[13] Lundell, G. E. F. and Knowles, H. B., *Journal of Research NBS*, Vol. 3, 1929, p. 92.
[14] Fleck, H. R. and War, A. M., *Analyst*, Vol. 58, 1933, p. 388.
[15] Goto, H., *Journal of the Chemical Society of Japan*, Vol. 54, 1933, p. 725.
[16] Knowles, H. B., *Journal Research NBS*, Vol. 15, 1935, p. 90.
[17] Berg, R., *Z. Anal. Chem.*, Vol. 70, p. 341.
[18] Roller, P. S., *Journal of American Chemical Society*, Vol. 55, 1933, pp. 2437–2438.

APPENDIX II
ASTM Methods for Chemical Analysis of Hydraulic Cement (C 114-85)

Designation: C 114 – 85

Standard Methods for
CHEMICAL ANALYSIS OF HYDRAULIC CEMENT[1]

This standard is issued under the fixed designation C 114; the number immediately following the designation indicates the year of original adoption or, in the case of revision, the year of last revision. A number in parentheses indicates the year of last reapproval. A superscript epsilon (ε) indicates an editorial change since the last revision or reapproval.

1. Scope

1.1 These methods cover the chemical analyses of hydraulic cements. Any methods of demonstrated acceptable precision and accuracy may be used for analysis of hydraulic cements, including analyses for referee and certification purposes, as explained in Section 3. Specific chemical methods are provided for ease of reference for those desiring to use them. They are grouped as Reference Methods and Alternate Methods. The reference methods are long accepted wet chemical methods which provide a reasonably well-integrated basic scheme of analysis for hydraulic cements. The alternate methods generally provide individual determination of specific components and may be used alone or as alternates and determinations within the basic scheme at the option of the analyst and as indicated in the individual method. The individual analyst must demonstrate achievement of acceptable precision and accuracy, as explained in Section 3, when these methods are used.

1.2 *Contents:*

Section	Subject
2	Applicable Documents
3	Number of Determinations and Permissible Variations
3.1	Referee Analyses
3.2	Optional Analyses
3.3	Performance Requirements for Rapid Methods
3.4	Precision and Accuracy
4	General
4.1	Interferences and Limitations
4.2	Apparatus and Materials
4.3	Reagents
4.4	Sample Preparation
4.5	General Procedures
4.6	Recommended Order for Reporting Analyses

Reference Methods

Section	Subject
5	Insoluble Residue
6	Silicon Dioxide
6.2	Cements with Insoluble Residue Less Than 1 %
6.3	Cements with Insoluble Residue Greater Than 1 %
7	Ammonium Hydroxide Group
8	Ferric Oxide
9	Phosphorus Pentoxide
10	Titanium Dioxide
11	Zinc Oxide
12	Aluminum Oxide
13	Calcium Oxide
14	Magnesium Oxide
15	Sulfur
15.1	Sulfur Trioxide
15.2	Sulfide
16	Loss On Ignition
16.1	Portland Cement
16.2	Portland Blast-Furnace Slag Cement and Slag Cement
17	Sodium and Potassium Oxides
17.1	Total Alkalis
17.2	Water-Soluble Alkalis
18	Manganic Oxide
19	Chloride
20	Chloroform-Soluble Organic Substances

Alternate Methods

Section	Subject
21	Calcium Oxide
22	Magnesium Oxide
23	Loss on Ignition
23.1	Portland Blast-Furnace Slag Cement and Slag Cement
24	Titanium Dioxide
25	Phosphorus Pentoxide
26	Manganic Oxide
27	Free Calcium Oxide

---

[1] These methods are under the jurisdiction of ASTM Committee C-1 on Cement and are the direct responsibility of Subcommittee C01.23 on Chemical Analysis.
Current edition approved July 16, 1985. Published September 1985. Originally published as C 114 – 34 T. Last previous edition C 114 – 83a.

1.3 The values stated in SI units are to be regarded as the standard.

1.4 *This standard may involve hazardous materials, operations, and equipment. This standard does not purport to address all of the safety problems associated with its use. It is the responsibility of whoever uses this standard to consult and establish appropriate safety and health practices and determine the applicability of regulatory limitations prior to use.* For specific precaution statements, see 11.1.2.

2. Applicable Documents

2.1 *ASTM Standards:*

C 115 Test Method for Fineness of Portland Cement by the Turbidimeter[2]

C 150 Specification for Portland Cement[2]

C 183 Methods of Sampling and Acceptance of Hydraulic Cement[2]

C 595 Specification for Blended Hydraulic Cements[2]

D 1193 Specification for Reagent Water[3]

E 29 Recommended Practice for Indicating Which Places of Figures Are to Be Considered Significant in Specified Limiting Values[4]

E 275 Practice for Describing and Measuring Performance of Ultraviolet, Visible, and Near Infrared Spectrophotometers[5]

E 617 Specification for Laboratory Weights and Precision Mass Standards[4]

E 832 Specification for Laboratory Filter Papers[4,6]

3. Number of Determinations and Permissible Variations

3.1 *Referee Analyses*—The reference methods that follow in Sections 5 through 20, or other methods qualified according to 3.3, are required for referee analysis in those cases where conformance to chemical specification requirements are questioned. In these cases, a cement shall not be rejected for failure to conform to chemical requirements unless all determinations of constituents involved and all necessary separations prior to the determination of any one constituent are made entirely by reference methods prescribed in the appropriate sections of this method or by other qualified methods, except when specific methods are prescribed in the standard specification for the cement in question. The methods actually used for the analysis shall be designated.

3.1.1 Referee analyses, when there is a question regarding acceptance, shall be made in duplicate and the analyses shall be made on different days. If the two results do not agree within the permissible variation given in Table 1, which appears at the end of this method, the determination shall be repeated until two or three results agree within the permissible variation. When two or three results do agree within the permissible variation, their average shall be accepted as the correct value. When an average of either two or three results can be calculated, the calculation shall be based on the three results. For the purpose of comparing analyses and calculating the average of acceptable results, the percentages shall be calculated to the nearest 0.01 (or 0.001 in the case of chloroform-soluble organic substances), although some of the average values are reported to 0.1 as indicated in the methods. When a blank determination is specified, one shall be made with each individual analysis or with each group of two or more samples analyzed on the same day for a given component.

3.1.2 Referee analyses or analyses intended for use as a basis for acceptance or rejection of a cement or for manufacturer's certification shall be made only after demonstration of precise and accurate analyses by the methods in use by meeting the requirements of 3.1.3, except when demonstrated under 3.3.2.1. Such demonstration may be made concurrently with analysis of the cement being tested and must have been made within the preceding two years. The demonstration is required only for those constituents being used as a basis for acceptance, rejection, or certification of a cement, but may be made for any constituent of cement for which a standard exists.

3.1.3 Demonstration shall be made by analysis of each constituent of concern in an NBS SRM cement (Note 1). Duplicate samples shall be run on different days. The same methods to be used for analysis of cement being tested shall be used for analysis of the NBS SRM cement. If the duplicate results do not agree within the permissible variation given in Table 1, the determinations shall be repeated, following identification and correction of problems or errors, until a set of duplicate results do agree within the permissible variation.

---

[2] *Annual Book of ASTM Standards*, Vol 04.01.
[3] *Annual Book of ASTM Standards*, Vol 11.01.
[4] *Annual Book of ASTM Standards*, Vol 14.02.
[5] *Annual Book of ASTM Standards*, Vol 14.01.
[6] *Annual Book of ASTM Standards*, Vol 15.09.

NOTE 1—The term SRM samples refers specifically to the seven NBS Standard Reference Materials Nos. 633 through 639.

3.1.4 The average of the results of acceptable duplicate determinations for each constituent may differ from the SRM certificate value by no more than the value shown in column 2 of Table 1 after correction for minor components when needed. When no SRM certificate value is given, a generally accepted accuracy standard for that constituent does not exist. In such cases, only the differences between duplicate values as specified in 3.1.3 shall apply.

3.1.5 Data demonstrating that precise and accurate results were obtained with NBS SRM cements by the same analyst making the acceptance determination shall be made available on request to all parties concerned when there is a question of acceptance of a cement.

3.2 *Optional Analyses*—The alternate methods provide, in some cases, procedures that are shorter or more convenient to use for routine determination of certain constituents than are the reference methods (Note 2). Longer, more complex procedures, in some instances, have been retained as alternate methods to permit comparison of results by different procedures or for use when unusual materials are being examined, where unusual interferences may be suspected, or when unusual preparation for analysis is required. Test results by alternate methods may be used as a basis for acceptance or rejection when it is clear that a cement does or does not meet the specification requirement.

NOTE 2—It is not intended that the use of reference methods be confined to referee analysis. A reference method may be used in preference to an alternate method when so desired. A reference method must be used where an alternate method is not provided.

3.2.1 Duplicate analyses and blank determinations are not required when using the alternate methods. If, however, a blank determination is desired for an alternate method, one may be used and it need not have been obtained concurrently with the analysis. The final results, when corrected for blank values, should, in either case, be so designated.

3.3 *Performance Requirements for Rapid Methods:*

3.3.1 *Definition and Scope*—Where analytical data obtained in accordance with this method are required, any method may be used that meets the requirements of 3.3.2. A method is considered to consist of the specific procedures, reagents, supplies, equipment, instrument, etc. selected and used in a consistent manner by a specific laboratory. See Note 3 for examples of procedures.

NOTE 3—Examples of methods used successfully by their authors for analysis of hydraulic cement are given in the list of references. Included are methods using atomic absorption X-ray spectrometry, and spectrophotometry-EDTA.

3.3.1.1 If more than one instrument, even though substantially identical, is used in a specific laboratory for the same analyses, use of each instrument shall constitute a separate method and each must be qualified separately.

3.3.2 *Qualification of a Method*—Prior to use for analysis of hydraulic cement, each method (see 3.3.1) must be qualified individually for such analysis. Qualification data, or if applicable, requalification data, shall be made available pursuant to the Manufacturer's Certification Section of Specification C 150 or the Certification Section of Specification C 595.

3.3.2.1 Using the method chosen, make single determinations for each oxide under consideration on each of the SRM samples (Note 1). Complete two rounds of tests on nonconsecutive days repeating all steps of sample preparations. Calculate the differences between values and averages of the values from the two rounds of tests.

3.3.2.2 At least six of the seven differences between duplicates obtained of any single component shall not exceed the limits shown in Column 2 of Table 1 and the remaining difference by no more than twice that value.

3.3.2.3 For each component and each SRM, the average obtained shall be compared to the certified concentrations. Where a certificate value includes a subscript number, that subscript shall be assumed to be a significant number. At least six of the seven averages for each component (oxide) shall not differ from the certified concentrations by more than the value shown in Column 3 of Table 1, and the remaining average by more than twice that value. The standardization, if needed, used for qualification and for analysis of each constituent shall be determined by valid curve-fitting procedures. The qualification testing shall be conducted with newly prepared specimens.

NOTE 4—Actual drawing of a curve is not required

if such curve is not needed for the method in use. A point-to-point, saw-tooth curve that is artificially made to fit a set of data points does not constitute a valid curve-fitting procedure.

3.3.3 *Partial Results*—Methods that provide acceptable results for some components but not for others may be used only for those components for which acceptable results are obtained.

3.3.4 *Report of Results*—Chemical analyses obtained by qualified rapid methods and reported pursuant to the Manufacturer's Certification Section of Specification C 150 or the Certification Section of Specification C 595 shall be indicated as having been obtained by rapid methods and the type of method used shall be designated.

3.3.5 *Rejection of Material*—See 3.1 and 3.2.

3.3.6 *Requalification of a Method:*

3.3.6.1 Requalification of a method, as defined in 3.3.2, shall be required at least every 2 years.

3.3.6.2 Requalification of a method also shall be required upon receipt of substantial evidence that the method may not be providing data in accordance with Table 1 for one or more constituents. Such requalification may be limited to those constituents indicated to be in error and shall be carried out prior to further use of the method for analysis of those constituents.

3.3.6.3 Substantial evidence that a method may not be providing data in accordance with Table 1 shall be considered to have been received when a laboratory is informed that analysis of the same material by Reference Methods run in accordance with 3.1.1, the final average of a CCRL sample, a certificate value of an NBS SRM, or an accepted value of a known secondary standard differs from the value obtained by the method in question by more than twice the value shown in Column 2 of Table 1 for one or more constituents. When indirect methods are involved, as when a value is obtained by difference, corrections shall be made for minor constituents in order to put analyses on a comparable basis prior to determining the differences. (See Note 5) For any constituents affected, a method also shall be requalified after any substantial repair or replacement of one or more critical components of an instrument essential to the method.

NOTE 5—Instrumental analyses can usually detect only the element sought. Therefore, to avoid controversy, the actual procedure used for the elemental analyses should be noted when actual differences with reference procedures can exist. For example, P_2O_5 and TiO_2 are included with Al_2O_3 in the usual wet method and sulfide sulfur is included in most instrumental procedures with SO_3.

3.3.6.4 If an instrument or piece of equipment is replaced, even if by one of identical make or model, or is significantly modified, a previously qualified method using such new or modified instrument or equipment shall be considered a new method and must be qualified in accordance with 3.3.2.

3.4 *Precision and Bias*—Different analytical methods are subject to individual limits of precision and bias. It is the responsibility of the user to demonstrate that the methods used at least meet the limits of precision and bias shown in Table 1.

4. General

4.1 *Interferences and Limitations:*

4.1.1 These methods were developed primarily for the analysis of portland cements. However, except for limitations noted in the procedure for specific constituents, the reference methods provide for accurate analyses of other hydraulic cements that are completely decomposed by hydrochloric acid, or where a preliminary sodium carbonate fusion is made to ensure complete solubility. Some of the alternate methods may not always provide accurate results because of interferences from elements which are not removed during the procedure.

4.1.2 When using a method that determines total sulfur, such as most instrumental methods, sulfide sulfur will be determined with sulfate and included as such. In most hydraulic cements, the difference resulting from such inclusion will be insignificant, less than 0.05 weight %. In some cases, notably slags and slag-containing cements but sometimes other cements as well, significant levels of sulfide may be present. In such cases, especially if there is a question of meeting or not meeting a specification limit or when the most accurate results are desired, analytical methods shall be chosen so that sulfate and sulfide can be reported separately.

4.2 *Apparatus and Materials:*

4.2.1 *Balance*—The analytical balance used in the chemical determinations shall conform to the following requirements:

4.2.1.1 The balance shall have a capacity of not more than 200 g. It may be of conventional design, either with or without "quick-weighing" devices, or it may be a constant-load, direct-reading type. It shall be capable of reproducing results within 0.0002 g with an accuracy of ± 0.0002 g. Direct-reading balances shall have a sensitivity not exceeding 0.0001 g (Note 6). Conventional two-pan balances shall have a maximum sensibility reciprocal of 0.0003 g. Any rapid weighing device that may be provided, such as a chain, damped motion, or heavy riders, shall not increase the basic inaccuracy by more than 0.0001 g at any reading and with any load within the rated capacity of the balance.

NOTE 6—The sensitivity of a direct-reading balance is the weight required to change the reading one graduation. The sensibility reciprocal for a conventional balance is defined as the change in weight required on either pan to change the position of equilibrium one division on the pointer scale at capacity or at any lesser load.

4.2.2 *Weights*—Weights used for analysis shall conform to Types I or II, Grades S or O, Classes 1, 2, or 3 as described in Specification E 617. They shall be checked at least once a year, or when questioned, and adjusted at least to within allowable tolerances for Class 3 weights (Note 7). For this purpose each laboratory shall also maintain, or have available for use, a reference set of standard weights from 50 g to 10 mg, which shall conform at least to Class 3 requirements and be calibrated at intervals not exceeding five years by the National Bureau of Standards. After initial calibration, recalibration by the National Bureau of Standards may be waived provided it can be shown by documented data obtained within the time interval specified that a weight comparison between summations of smaller weights and a single larger weight nominally equal to that summation, establishes that the allowable tolerances have not been exceeded. All new sets of weights purchased shall have the weights of 1 g and larger made of stainless steel or other corrosion-resisting alloy not requiring protective coating, and shall meet the density requirements for Grades S or O.

NOTE 7—The scientific supply houses do not presently list weights as meeting Specification E 617. They list weights as meeting NBS or OIML standards. The situation with regard to weights is in a state of flux because of the trend toward internationalization. Hopefully this will soon be resolved.

NBS Classes S and S-1 and OIML Class F_1 weights meet the requirements of this standard.

4.2.3 *Glassware and Laboratory Containers*—Standard volumetric flasks, burets, and pipets should be of precision grade or better. Standard-taper, interchangeable, ground-glass joints are recommended for all volumetric glassware and distilling apparatus, when available. Wherever applicable, the use of special types of glassware, such as colored glass for the protection of solutions against light, alkali-resistant glass, and high-silica glass having exceptional resistance to thermal shock is recommended. Polyethylene containers are recommended for all aqueous solutions of alkalies and for standard solutions where the presence of dissolved silica or alkali from the glass would be objectionable. Such containers shall be made of high-density polyethylene having a wall thickness of at least 1 mm.

4.2.4 *Desiccators*—Desiccators shall be provided with a good desiccant, such as magnesium perchlorate, activated alumina, or sulfuric acid. Anhydrous calcium sulfate may also be used provided it has been treated with a color-change indicator to show when it has lost its effectiveness. Calcium chloride is not a satisfactory desiccant for this type of analysis.

4.2.5 *Filter Paper*—Filter paper shall conform to the requirements of Specification E 832, Type II, Quantitative. When coarse-textured paper is required, Class E paper shall be used, when medium-textured paper is required, Class F paper shall be used, and when retentive paper is required, Class G shall be used.

4.2.6 *Crucibles*—Platinum crucibles for ordinary chemical analysis should preferably be made of pure unalloyed platinum and be of 15 to 30-mL capacity. Where alloyed platinum is used for greater stiffness or to obviate sticking of crucible and lid, the alloyed platinum should not decrease in weight by more than 0.2 mg when heated at 1200°C for 1 h.

4.2.7 *Muffle Furnace*—The muffle furnace shall be capable of operation at the temperatures required and shall have an indicating pyrometer accurate within ±25°C, as corrected, if necessary, by calibration. More than one furnace may be used provided each is used within its proper operating temperature range.

4.3 *Reagents:*

4.3.1 *Purity of Reagents*—Reagent grade chemicals shall be used in all tests. Unless otherwise indicated, it is intended that all reagents shall conform to the specifications of the Committee on Analytical Reagents of the American Chemical Society, where such specifications are available.[7] Other grades may be used, provided it is first ascertained that the reagent is of sufficiently high purity to permit its use without lessening the accuracy of the determination.

4.3.2 Unless otherwise indicated, references to water shall be understood to mean reagent water conforming to Specification D 1193.

4.3.3 *Concentration of Reagents:*

4.3.3.1 *Prepackaged Reagents*—Commercial prepackaged standard solutions or diluted prepackaged concentrations of a reagent may be used whenever that reagent is called for in the procedures provided that the purity and concentrations are as specified. Verify purity and concentration of such reagents by suitable tests.

4.3.3.2 *Concentrated Acids and Ammonium Hydroxide*—When acids and ammonium hydroxide are specified by name or chemical formula only, it shall be understood that concentrated reagents of the following specific gravities or concentrations by weight are intended:

Acetic acid ($HC_2H_3O_2$)	99.5 %
Hydrochloric acid (HCl)	sp gr 1.19
Hydrofluoric acid (HF)	48 %
Nitric acid (HNO_3)	sp gr 1.42
Phosphoric acid (H_3PO_4)	85 %
Sulfuric acid (H_2SO_4)	sp gr 1.84
Ammonium hydroxide (NH_4OH)	sp gr 0.90

4.3.3.3 The desired specific gravities or concentrations of all other concentrated acids shall be stated whenever they are specified.

4.3.4 *Diluted Acids and Ammonium Hydroxide*—Concentrations of diluted acids and ammonium hydroxide, except when standardized, are specified as a ratio stating the number of volumes of the concentrated reagent to be added to a given number of volumes of water, for example: HCl (1+99) means 1 volume of concentrated HCl (sp gr 1.19) added to 99 volumes of water.

4.3.5 *Standard Solutions*—Concentrations of standard solutions shall be expressed as normalities (N) or as equivalents in grams per millilitre of the component to be determined, for example: 0.1 N $Na_2S_2O_3$ solution or $K_2Cr_2O_7$ (1 mL = 0.004 g Fe_2O_3). The average of at least three determinations shall be used for all standardizations. When a material is used as a primary standard, reference has generally been made to the standard furnished by the National Bureau of Standards. However, when primary standard grade materials are otherwise available they may be used or the purity of a salt may be determined by suitable tests.

4.3.6 *Nonstandardized Solutions*—Concentrations of nonstandardized solutions prepared by dissolving a given weight of the solid reagent in a solvent shall be specified in grams of the reagent per litre of solution, and it shall be understood that water is the solvent unless otherwise specified, for example: NaOH solution (10 g/L) means 10 g of NaOH dissolved in water and diluted with water to 1 L. Other nonstandardized solutions may be specified by name only, and the concentration of such solutions will be governed by the instructions for their preparation.

4.3.7 *Indicator Solutions:*

4.3.7.1 *Methyl Red*—Prepare the solution on the basis of 2 g of methyl red/litre of 95 % ethyl alcohol.

4.3.7.2 *Phenolphthalein*—Prepare the solution on the basis of 1 g of phenolphthalein per litre of 95 % ethyl alcohol.

4.4 *Sample Preparation:*

4.4.1 Before testing, pass representative portions of each sample through a No. 20 (850-μm) sieve, or any other sieve having approximately 20 openings per inch, in order to mix the sample, break up lumps, and remove foreign materials. Discard the foreign materials and hardened lumps that do not break up on sieving or brushing.

4.4.2 By means of a sample splitter or by quartering, the representative sample shall be reduced to a laboratory sample of at least 50 g. Where larger quantities are required for additional determinations such as water-soluble alkali, chloride, duplicate testing, etc., prepare a sample of at least 100 g.

4.4.3 Pass the laboratory sample through a U. S. No. 100 sieve (sieve opening of 150 μm). Further grind the sieve residue so that it also passes the No. 100 sieve. Homogenize the entire sample by again passing it through the sieve.

4.4.4 Transfer the sample to a clean, dry, glass container with an airtight lid and further mix the sample thoroughly.

4.4.5 Expedite the above procedure so that the sample is exposed to the atmosphere for a

---

[7] Reagent Chemicals, American Chemical Society Specifications," Am. Chemical Soc., Washington, DC. For suggestions on the testing of reagents not listed by the American Chemical Society, see "Reagent Chemicals and Standards," by Joseph Rosin, D. Van Nostrand Co., Inc., New York, NY, and the "United States Pharmacopeia."

minimum time.

4.5 *General Procedures:*

4.5.1 *Weighing*—The calculations included in the individual procedures assume that the exact weight specified has been used. Accurately weighed samples, which are approximately but not exactly equal to the weight specified, may be used provided appropriate corrections are made in the calculations. Unless otherwise stated, weights of all samples and residues should be recorded to the nearest 0.0001 g.

4.5.2 *Tared or Weighed Crucibles*—The tare weight of crucibles shall be determined by preheating the empty crucible to constant weight at the same temperature and under the same conditions as shall be used for the final ignition of a residue and cooling in a desiccator for the same period of time used for the crucible containing the residue.

4.5.3 *Constancy of Weight of Ignited Residues*—To definitely establish the constancy of weight of an ignited residue for referee purposes, the residue shall be ignited at the specified temperature and for the specified time, cooled to room temperature in a desiccator, and weighed. The residue shall then be reheated for at least 30 min, cooled to room temperature in a desiccator, and reweighed. If the two weights do not differ by more than 0.2 mg, constant weight is considered to have been attained. If the difference in weights is greater than 0.2 mg, additional ignition periods are required until two consecutive weights agree within the specified limits. For ignition loss, each reheating period shall be 5 min.

4.5.4 *Volatilization of Platinum*—The possibility of volatilization of platinum or alloying constituents from the crucibles must be considered. On reheating, if the crucible and residue lose the same weight (within 0.2 mg) as the crucible containing the blank, constant weight can be assumed. Crucibles of the same size, composition, and history shall be used for both the sample and the blank.

4.5.5 *Calculations*—In all operations on a set of observed values such as multiplying or dividing, where possible, retain the equivalent of two more places of figures than in the single observed values. For example, if observed values are read or determined to the nearest 0.1 mg, carry numbers to the nearest 0.001 mg in calculation.

4.5.6 *Rounding Figures*—Rounding of figures to the number of significant places required in the report should be done after calculations are completed, in order to keep the final results substantially free of calculation errors. The rounding procedure should follow the principle outlined in Recommended Practice E 29.[8]

NOTE 8—The rounding procedure referred to in 4.5.6, in effect, drops all digits beyond the number of places to be retained if the next figure is less than 5. If it is more than 5, or equal to 5 and subsequent places contain a digit other than 0, then the last retained digit is increased by one. When the next digit is equal to 5 and all other subsequent digits are 0, the last digit to be retained is unchanged when it is even and increased by one when it is odd. For example 3.96 (50) remains 3.96 but 3.95 (50) becomes 3.96.

4.6 *Recommended Order for Reporting Analyses*—The following order is recommended for reporting the results of chemical analysis of portland cement:

Major Components:
 SiO_2 (silicon dioxide)
 Al_2O_3 (aluminum oxide)
 Fe_2O_3 (ferric oxide)
 CaO (calcium oxide)
 MgO (magnesium oxide)
 SO_3 (sulfur trioxide)
 Loss on ignition
Minor Components:
 Na_2O (sodium oxide)
 K_2O (potassium oxide)
 TiO_2 (titanium dioxide)
 P_2O_5 (phosphorus pentoxide)
 ZnO (zinc oxide)
 Mn_2O_3 (manganic oxide)
 Sulfide sulfur
Separate Determinations:
 Insoluble residue
 Chloroform-soluble organic substances
 Free calcium oxide
 Water-soluble alkali

REFERENCE METHODS

5. Insoluble Residue (*Reference Method*)

5.1 *Summary of Method:*

5.1.1 In this procedure, insoluble residue of a cement is determined by digestion of the sample in hydrochloric acid followed, after filtration, by further digestion in sodium hydroxide. The resulting residue is ignited and weighed (Note 9).

NOTE 9—This method, or any other method designed for the estimation of an acid-insoluble substance in any type of cement, is empirical because the amount obtained depends on the reagents and the time and temperature of digestion. If the amount is large, there may be a little variation in duplicate determinations.

---

[8] See also the *ASTM Manual on Presentation of Data and Control Chart Analysis*, STP 15D, 1976.

The procedure should be followed closely in order to reduce the variation to a minimum.

5.1.2 When the method is used on blended cement, the decomposition in acid is considered to be complete when the portland cement clinker is decomposed completely. An ammonium nitrate solution is used in the final washing to prevent finely ground insoluble material from passing through the filter paper.

5.2 *Reagents:*

5.2.1 *Ammonium Nitrate Solution* (20 g NH_4NO_3/L).

5.2.2 *Sodium Hydroxide Solution* (10 g NaOH/L).

5.3 *Procedure:*

5.3.1 To 1 g of the sample (Note 10) add 25 mL of cold water. Disperse the cement in the water and while swirling the mixture, quickly add 5 mL of HCl. If necessary, warm the solution gently, and grind the material with the flattened end of a glass rod for a few minutes until it is evident that decomposition of the cement is complete (Note 11). Dilute the solution to 50 mL with hot water (nearly boiling) and heat the covered mixture rapidly to near boiling by means of a high-temperature hot plate. Then digest the covered mixture for 15 min at a temperature just below boiling (Note 12). Filter the solution through a medium-textured paper into a 400-mL beaker, wash the beaker, paper, and residue thoroughly with hot water, and reserve the filtrate for the sulfur trioxide determination, if desired (Note 13). Transfer the filter paper and contents to the original beaker, add 100 mL of hot (near boiling) NaOH solution (10 g/L), and digest at a temperature just below boiling for 15 min. During the digestion, occasionally stir the mixture and macerate the filter paper. Acidify the solution with HCl using methyl red as the indicator and add an excess of 4 or 5 drops of HCl. Filter through medium-textured paper and wash the residue at least 14 times with hot NH_4NO_3 solution (20 g/L) making certain to wash the entire filter paper and contents during each washing. Ignite the residue in a weighed platinum crucible at 900 to 1000°C, cool in a desiccator, and weigh.

NOTE 10—If sulfur trioxide is to be determined by turbidimetry it is permissible to determine the insoluble residue on a 0.5-g sample. In this event, the percentage of insoluble residue should be calculated to the nearest 0.01 by multiplying the weight of residue obtained by 200. However, the cement should not be rejected for failure to meet the insoluble residue requirement unless a 1-g sample has been used.

NOTE 11—If a sample of portland cement contains an appreciable amount of manganic oxide, there may be brown compounds of manganese which dissolve slowly in cold diluted HCl but rapidly in hot HCl in the specified strength. In all cases, dilute the solution as soon as decomposition is complete.

NOTE 12—In order to keep the solutions closer to the boiling temperature, it is recommended that these digestions be carried out on an electric hot plate rather than on a steam bath.

NOTE 13—Continue with the sulfur trioxide determination (15.1.2.1 through 15.1.3) by diluting to 250 or 200 mL as required by the appropriate section.

5.3.2 *Blank*—Make a blank determination, following the same procedure and using the same amounts of reagents, and correct the results obtained in the analysis accordingly.

5.4 *Calculation*—Calculate the percentage of the insoluble residue to the nearest 0.01 by multiplying the weight in grams of the residue (corrected for the blank) by 100.

6. Silicon Dioxide (*Reference Method*)

6.1 *Selection of Method*—For cements other than portland and for which the insoluble residue is unknown, determine the insoluble residue in accordance with Section 5 of these methods. For portland cements and other cements having an insoluble residue less than 1 %, proceed in accordance with 6.2. For cements having an insoluble residue greater than 1 % proceed in accordance with 6.3.

6.2 *Silicon Dioxide in Portland Cements and Cements with Low Insoluble Residue:*

6.2.1 *Summary of Method*—In this method silicon dioxide (SiO_2) is determined gravimetrically. Ammonium chloride is added and the solution is not evaporated to dryness. This method was developed primarily for hydraulic cements that are almost completely decomposed by hydrochloric acid and should not be used for hydraulic cements that contain large amounts of acid-insoluble material and require a preliminary sodium carbonate fusion. For such cements, or if prescribed in the standard specification for the cement being analyzed, the more lengthy procedure in 6.3 shall be used.

6.2.2 *Reagent*—Ammonium chloride (NH_4Cl).

6.2.3 *Procedure:*

6.2.3.1 Mix thoroughly 0.5 g of the sample and about 0.5 g of NH_4Cl in a 50-mL beaker, cover the beaker with a watch glass, and add cautiously 5 mL of HCl, allowing the acid to run down the lip of the covered beaker. After the chemical action has subsided, lift the cover, add 1 or 2 drops of HNO_3, stir the mixture with a glass rod, replace the cover, and set the beaker on a steam bath for 30 min (Note 14). During this time of digestion, stir the contents occasionally and break up any remaining lumps to facilitate the complete decomposition of the cement. Fit a medium-textured filter paper to a funnel, transfer the jelly-like mass of silicic acid to the filter as completely as possible without dilution, and allow the solution to drain through. Scrub the beaker with a policeman and rinse the beaker and policeman with hot HCl (1+99). Wash the filter two or three times with hot HCl (1+99) and then with ten or twelve small portions of hot water, allowing each portion to drain through completely. Reserve the filtrate and washings for the determination of the ammonium hydroxide group (Note 15).

NOTE 14—A hot plate may be used instead of a steam bath if the heat is so regulated as to approximate that of a steam bath.
Under conditions where water boils at a lower temperature than at sea level: such as at higher elevations, 30 min may not be sufficient to recover all of the silica. In such cases, increase the time of digestion as necessary to get complete recovery of the silica. In no case should this time exceed 60 min.

NOTE 15—Determine the ammonium hydroxide group in accordance with the procedure described in 7.1 through 7.3.

6.2.3.2 Transfer the filter paper and residue to a weighed platinum crucible, dry, and ignite, at first slowly until the carbon of the paper is completely consumed without inflaming, and finally at 1100 to 1200°C for 1 h. Cool in a desiccator and weigh. Reignite to constant weight. Treat the SiO_2 thus obtained, which will contain small amounts of impurities, in the crucible with 1 or 2 mL of water, 2 drops of H_2SO_4 (1+1), and about 10 mL of HF, and evaporate cautiously to dryness. Finally, heat the small residue at 1050 to 1100°C for 5 min, cool in a desiccator, and weigh. The difference between this weight and the weight previously obtained represents the weight of SiO_2. Consider the weighed residue remaining after the volatilization of SiO_2 as combined aluminum and ferric oxides and add it to the result obtained in the determination of the ammonium hydroxide group.

6.2.3.3 If the HF residue exceeds 0.0020 g, the silica determination shall be repeated, steps should be taken to ensure complete decomposition of the sample before a silica separation is attempted, and the balance of the analysis (ammonium hydroxide group, CaO, and MgO) determined on the new silica filtrate provided the new silica determination has a HF residue of 0.0020 g or less except as provided in 6.2.3.4 and 6.2.3.5.

6.2.3.4 If two or three repeated determinations of a sample of portland cement consistently show HF residues higher than 0.0020 g, this is evidence that contamination has occurred in sampling or the cement has not been burned properly during manufacture. In such a case, do not fuse the large HF residue with pyrosulfate for subsequent addition to the filtrate from the silica separation. Instead, report the value obtained for the HF residue. Do not ignite the ammonium hydroxide group in the crucible conta·ning this abnormally large HF residue.

6.2.3.5 In the analysis of cements other than portland, it may not always be possible to obtain HF residues under 0.0020 g. In such cases, add 0.5 g of sodium or potassium pyrosulfate ($Na_2S_2O_7$ or $K_2S_2O_7$) to the crucible and heat below red heat until the small residue of impurities is dissolved in the melt (Note 16). Cool, dissolve the fused mass in water, and add it to the filtrate and washings reserved for the determination of the ammonium hydroxide group.

NOTE 16—A supply of nonspattering pyrosulfate may be prepared by heating some pyrosulfate in a platinum vessel below red heat until the foaming and spattering cease, cooling, and crushing the fused mass.

6.2.3.6 *Blank*—Make a blank determination, following the same procedure and using the same amounts of reagents, and correct the results obtained in the analysis accordingly.

6.2.4 *Calculation*—Calculate the percentage of SiO_2 to the nearest 0.1 multiplying the weight in grams of SiO_2 by 200 (100 divided by the weight of the sample used (0.5 g)).

6.3 *Silicon Dioxide in Cements with Insoluble Residue Greater Than 1 %:*

6.3.1 *Summary of Method*—This method is based on the sodium carbonate fusion followed by double evaporation to dryness of the hydrochloric acid solution of the fusion product to convert silicon dioxide (SiO_2) to the insoluble form. The solution is filtered and the insoluble siliceous residue is ignited and weighed. Silicon dioxide is volatilized by hydrofluoric acid and the loss of weight is reported as pure SiO_2.

6.3.2 *Procedure:*

6.3.2.1 Weigh a quantity of the ignited sample equivalent to 0.5 g of the as-received sample calculated as follows:

$$W = [(0.5 (100.00 - I)]/100$$

where:

W = weight of ignited sample, g, and
I = loss of ignition, %.

The ignited material from the loss on ignition determination may be used for the sample. Thoroughly mix the sample with 4 to 6 g of Na_2CO_3 by grinding in an agate mortar. Place a thin layer of Na_2CO_3 on the bottom of a platinum crucible of 20 to 30-mL capacity, add the cement-Na_2CO_3 mixture, and cover the mixture with a thin layer of Na_2CO_3. Place the covered crucible over a moderately low flame and increase the flame gradually to a maximum (approximately 1100°C) and maintain this temperature until the mass is quiescent (about 45 min). Remove the burner, lay aside the cover of the crucible, grasp the crucible with tongs, and slowly rotate the crucible so that the molten contents spread over the sides and solidify as a thin shell on the interior. Set the crucible and cover aside to cool. Rinse off the outside of the crucible and place the crucible on its side in a 300-mL casserole about one third full of water. Warm the casserole and stir until the cake in the crucible disintegrates and can be removed easily. By means of a glass rod, lift the crucible out of the liquid, rinsing it thoroughly with water. Rinse the cover and crucible with HCl (1+3); then add the rinse to the casserole. Very slowly and cautiously add 20 mL of HCl (sp gr 1.19) to the covered casserole. Remove the cover and rinse. If any gritty particles are present, the fusion is incomplete and the test must be repeated, using a new sample. **Caution:** Subsequent steps of the methods must be followed exactly for accurate results.

6.3.2.2 Evaporate the solution to dryness on a steam bath (there is no longer a gelatinous appearance). Without heating the residue any further, treat it with 5 to 10 mL of HCl, wait at least 2 min, and then add an equal amount of water. Cover the dish and digest for 10 min on the steam bath or a hot plate. Dilute the solution with an equal volume of hot water, immediately filter through medium-textured paper and wash the separated SiO_2 thoroughly with hot HCl (1+99), then with hot water. Reserve the residue.

6.3.2.3 Again evaporate the filtrate to dryness, and bake the residue in an oven for 1 h at 105 to 110°C. Cool, add 10 to 15 mL of HCl (1+1), and digest on the steam bath or hot plate for 10 min. Dilute with an equal volume of water, filter immediately on a fresh filter paper, and wash the small SiO_2 residue thoroughly as described in 6.3.2.2. Stir the filtrate and washings and reserve for the determination of the ammonium hydroxide group in accordance with 7.1 through 7.3.

6.3.2.4 Continue the determination of silicon dioxide in accordance with 6.2.3.2.

7. Ammonium Hydroxide Group (*Reference Method*)

7.1 *Summary of Method*—In this method aluminum, iron, titanium, and phosphorus are precipitated from the filtrate, after SiO_2 removal, by means of ammonium hydroxide. With care, little if any manganese will be precipitated. The precipitate is ignited and weighed as the oxides.

7.2 *Procedure:*

7.2.1 To the filtrate reserved in accordance with 6.2.3.1 (Note 17) which should have a volume of about 200 mL, add HCl if necessary to ensure a total of 10 to 15 mL of the acid. Add a few drops of methyl red indicator and heat to boiling. Then treat with NH_4OH (1+1) (Note 18), dropwise until the color of the solution becomes distinctly yellow, and add one drop in excess (Note 19). Heat the solution containing the precipitate to boiling and boil for 50 to 60 s. In the event difficulty from bumping is experienced while boiling the ammoniacal solution, a digestion period of 10 min on a steam bath, or on a hot plate having the approximate temperature of a steam bath, may be substituted for the 50 to 60-s boiling period. Allow the precipitate to settle (not more than 5 min) and filter using medium-textured paper (Note 20). Wash, from twice for a small precipitate to four times for a large one, with hot ammonium nitrate (NH_4NO_3, 20 g/L) (Note 21).

NOTE 17—If a platinum evaporating dish has been used for the dehydration of SiO_2, iron may have been partially reduced. At this stage, add about 3 mL of saturated bromine water to the filtrate and boil the filtrate to eliminate the excess bromine before adding the methyl red indicator. If difficulty from bumping is experienced during the boiling, the following alternate techniques may be helpful: (*1*) a piece of filter paper, approximately 1 cm² in area, positioned where the bottom and side of the beaker merge and held down by the end of a stirring rod may solve the difficulty, and

(2) use of 400-mL beakers supported inside a cast aluminum cup has also been found effective.

NOTE 18—The NH_4OH used to precipitate the hydroxides must be free of contamination with carbon dioxide (CO_2).

NOTE 19—It usually takes 1 drop of NH_4OH (1+1) to change the color of the solution from red to orange and another drop to change the color from orange to yellow. If desired, the addition of the indicator may be delayed until ferric hydroxide ($Fe(OH)_3$) is precipitated without aluminum hydroxide ($Al(OH)_3$) being completely precipitated. In such a case, the color changes may be better observed. However, if the content of Fe_2O_3 is unusually great, it may be necessary to occasionally let the precipitate settle slightly so that the color of the supernatant liquid can be observed. If the color fades during the precipitation, add more of the indicator. Observation of the color where a drop of the indicator strikes the solution may be an aid in the control of the acidity. The boiling should not be prolonged as the color may reverse and the precipitate may be difficult to retain on the filter. The solution should be distinctly yellow when it is ready to filter. If it is not, restore the yellow color with more NH_4OH (1+1) or repeat the precipitation.

NOTE 20—To avoid drying of the precipitate with resultant slow filtration, channeling, or poor washing, the filter paper should be kept nearly full during the filtration and should be washed without delay.

NOTE 21—Two drops of methyl red indicator solution should be added to the NH_4NO_3 solution in the wash bottle, followed by NH_4OH (1+1) added dropwise until the color just changes to yellow. If the color reverts to red at any time due to heating, it should be brought back to yellow by the addition of a drop of NH_4OH (1+1).

7.2.2 Set aside the filtrate and transfer the precipitate and filter paper to the same beaker in which the first precipitation was effected. Dissolve the precipitate with hot HCl (1+2). Stir to thoroughly macerate the paper and then dilute the solution to about 100 mL. Reprecipitate the hydroxides as described in 7.2.1. If difficulty from bumping is experienced while boiling the acid solution containing the filter paper, it may be obviated by diluting the hot 1+2 solution of the mixed oxides with 100 mL of boiling water and thus eliminate the need for boiling. Filter the solution and wash the precipitate with about four 10-mL portions of hot NH_4NO_3 solution (20 g/L) (Note 21). Combine the filtrate and washings with the filtrate set aside and reserve for the determination of CaO in accordance with 13.3.1.

7.2.3 Place the precipitate in a weighed platinum crucible, heat slowly until the papers are charred, and finally ignite to constant weight at 1050 to 1100°C taking care to prevent reduction, and weigh as the ammonium hydroxide group.

7.2.4 *Blank*—Make a blank determination, following the same procedure and using the same amounts of reagents, and correct the results obtained in the analysis accordingly.

7.3 *Calculation*—Calculate the percentage of ammonium hydroxide group to the nearest 0.01 by multiplying the weight in grams of ammonium hydroxide group by 200 (100 divided by the weight of sample used (0.5 g)).

8. Ferric Oxide (*Reference Method*)

8.1 *Summary of Method*—In this method, the Fe_2O_3 content of the cement is determined on a separate portion of the cement by reducing the iron to the ferrous state with stannous chloride ($SnCl_2$) and titrating with a standard solution of potassium dichromate ($K_2Cr_2O_7$). This determination is not affected by any titanium or vanadium that may be present in the cement.

8.2 *Reagents:*

8.2.1 *Barium Diphenylamine Sulfonate Indicator Solution*—Dissolve 0.3 g of barium diphenylamine sulfonate in 100 mL of water.

8.2.2 *Potassium Dichromate, Standard Solution* (1 mL = 0.004 g Fe_2O_3)—Pulverize and dry primary standard potassium dichromate ($K_2Cr_2O_7$) reagent, the current lot of NBS 136, at 180 to 200°C to constant weight. Weigh accurately an amount of dried reagent equal to 2.45700 g times the number of litres of solution to be prepared. Dissolve in water and dilute to exactly the required volume in a single volumetric flask of the proper size. This solution is a primary standard and requires no further standardization.

NOTE 22—Where large quantities of standard solution are required, it may be desirable for certain laboratories to use commercially produced primary standard potassium dichromate for most determinations. Such a material may be used provided that the first solution made from the container is checked, as follows: Using a standard solution of NBS 136, prepared as described in 8.2.2, analyze, in duplicate, samples of a NBS SRM cement (see Note 1), by the procedure given in 8.3.1.3 and 8.3.1.4. Repeat using a similar solution prepared from the commercial primary standard dichromate. The average percentages of Fe_2O_3 found by each method should not differ by more than 0.06 %.

8.2.3 *Stannous Chloride Solution*—Dissolve 5 g of stannous chloride ($SnCl_2 \cdot 2H_2O$) in 10 mL of HCl and dilute to 100 mL. Add scraps of iron-free granulated tin and boil until the solution is clear. Keep the solution in a closed dropping bottle containing metallic tin.

8.3 *Procedure*—For cements other than portland and for which the insoluble residue is unknown, determine the insoluble residue in accordance with the appropriate sections of these methods. When insoluble residue is known, proceed in accordance with 8.3.1 or 8.3.2 as is appropriate for the cement being analyzed.

8.3.1 For portland cements and cements having insoluble residue lower than 1 %, weigh 1 g of the sample into a 500-mL Phillips beaker or other suitable container. Add 40 mL of cold water and, while the beaker is being swirled, add 10 mL of HCl. If necessary, heat the solution and grind the cement with the flattened end of a glass rod until it is evident that the cement is completely decomposed. Continue the analysis in accordance with 8.3.3.

8.3.2 For cements with insoluble residue greater than 1 %, disperse 1 g of the sample in 10 mL of water in a 250-mL beaker. Digest with 10 mL of HCl and evaporate to dryness. Cool the beaker and add 10 mL of HCl, bring to boiling, add 20 mL of water, boil, and filter through a medium-textured paper folded inside a retentive paper. Wash the papers and contents six times with hot water and then return the filtrates to the original beaker and reserve. Place the papers and HCl-insoluble contents into a platinum crucible, burn off the paper, and ignite at 950 to 1050°C for 10 min. Cool the crucible, add 3 mL of dilute sulfuric acid (H_2SO_4, 1+1) and about 1 mL of HF. Heat gently and after the actual reaction has abated somewhat, add HF until the crucible is two thirds full. Evaporate slowly until white fumes are observed, add 2 mL of dilute H_2SO_4 (1+1) and about 1 mL of HF, and evaporate to dense fumes. Avoid prolonged fuming. Cool the crucible and contents and place into the beaker containing the reserved filtrate. Heat until the contents of the crucible are dissolved. Wash the crucible while removing from the beaker. Add 5 mL of HCl to the contents of the beaker. Concentrate the solution, if necessary, to about 125 mL. Continue the analysis in accordance with 8.3.3.

8.3.3 Heat the solution to boiling and treat it with the $SnCl_2$ solution, added dropwise while stirring and boiling, until the solution is decolorized. Add 1 drop in excess and cool the solution to room temperature by placing the beaker in a pan of cool water. After cooling and without delay, rinse the inside of the vessel with water, and add all at once 10 mL of a cool, saturated mercuric chloride ($HgCl_2$) solution. Stir the solution vigorously for 1 min by swirling the beaker and add 10 mL of H_3PO_4 (1+1) and 2 drops of barium diphenylamine sulfonate indicator. Add sufficient water so that the volume after titration will be between 75 and 100 mL. Titrate with the standard $K_2Cr_2O_7$ solution. The end point shall be taken as the point at which a single drop causes an intense purple coloration that remains unchanged on further addition of standard $K_2Cr_2O_7$ solution.

8.3.4 *Blank*—Make a blank determination following the same procedure and using the same amounts of reagents. Record the volume of $K_2Cr_2O_7$ solution required to establish the end point as described in 8.3.3. As some iron must be present to obtain the normal end point, if no definite purple color is obtained after the addition of 4 drops of the standard $K_2Cr_2O_7$ solution, record the blank as zero.

8.4 *Calculation:*

8.4.1 Calculate the percentage of Fe_2O_3 to the nearest 0.01 (to be reported to the nearest 0.1) as follows:

$$Fe_2O_3, \% = E(V - B) \times 100$$

where:

E = Fe_2O_3 equivalent of the $K_2Cr_2O_7$ solution, g/mL,
V = millilitres of $K_2Cr_2O_7$ solution required by the 1-g sample used, and
B = millilitres of $K_2Cr_2O_7$ solution required by the blank determination.

9. Phosphorus Pentoxide (*Reference Method*)

9.1 *Summary of Method*—This colorimetric method is applicable to the determination of P_2O_5 in portland cement. Under the conditions of the test, no constituent normally present in portland cement will interfere.

9.2 *Apparatus:*

9.2.1 *Spectrophotometer* (Note 23):

9.2.1.1 The instrument shall be equipped to measure absorbance of solutions at a spectral wavelength of 725 nm.

9.2.1.2 Wavelength measurements shall be repeatable within ±1 nm or less.

9.2.1.3 In the absorbance range from 0.1 to 1.0, the absorbance measurements shall be repeatable within ±1 % or less.

9.2.1.4 To establish that the spectrophotometer will permit a satisfactory degree of accuracy, qualify the instrument in accordance with 3.3.2 using the procedure in 9.4.1 through 9.4.9.

NOTE 23—For the measurement of the performance of the spectrophotometer, refer to Practice E 275.

9.3 *Reagents:*

9.3.1 *Ammonium Molybdate Solution*—Into a 1-L volumetric flask introduce 500.0 mL of 10.6 N H_2SO_4 (9.3.7). Dissolve 25.0 g of ammonium molybdate $((NH_4)_6MO_7O_{24} \cdot 4H_2O)$ in about 250 mL of warm water and transfer to the flask containing the H_2SO_4, while swirling the flask. Cool, dilute to 1 L with water, and store in a plastic bottle.

9.3.2 *Ascorbic Acid Powder*—For ease in dissolving, the finest mesh available should be used.

9.3.3 *Hydrochloric Acid, Standard* (6.5 ± 0.1 N)—Dilute 540 mL of concentrated HCl (sp gr 1.19) to 1 L with water. Standardize against standard NaOH solution (9.3.6) using phenolphthalein as indicator. Determine the exact normality and adjust to 6.5 ± 0.1 N by dilution with water. Restandardize to ensure that the proper normality has been achieved.

9.3.4 *Phosphate, Standard Solution A*—Dissolve 0.1917 g of oven-dried potassium dihydrogen phosphate (KH_2PO_4) in water and dilute to 1 L in a volumetric flask.

9.3.5 *Phosphate, Standard Solution B*—Dilute 50.0 mL of phosphate solution A to 500 mL with water.

9.3.6 *Sodium Hydroxide, Standard Solution* (1 N)—Dissolve 40.0 g of sodium hydroxide (NaOH) in water, add 10 mL of a freshly filtered saturated solution of barium hydroxide $(Ba(OH)_2)$, and dilute to 1 L with water that has been recently boiled and cooled. Shake the solution from time to time during a several-hour period, and filter into a plastic bottle. Keep the bottle tightly closed to protect the solution from CO_2 in the air. Standardize against acid potassium phthalate or benzoic acid acidimetric standards furnished by the National Bureau of Standards (standard samples 84f and 350), using the methods in the certificates accompanying the standard samples. Determine the exact normality of the solution.

9.3.7 *Sulfuric Acid, Standard* (10.6 ± 0.1 N)— To a 1-L volumetric flask cooled in water add about 600 mL of water and then, slowly, with *caution*, 300 mL of concentrated H_2SO_4 (sp gr 1.84). After cooling to room temperature, dilute to 1 L with water. Standardize against the standard NaOH solution (9.3.6) using phenolphthalein as indicator. Determine the normality and adjust to 10.6 ± 0.1 N by dilution with water. Restandardize to ensure that the proper normality has been achieved.

9.4 *Procedure:*

9.4.1 Prepare a series of phosphate solutions to cover the range from 0 to 0.5 % P_2O_5. Prepare each solution by adding a suitable volume of standard phosphate solution B and 25.0 mL of the 6.5 N hydrochloric acid to a 250-mL volumetric flask (Note 24). Dilute to the mark with water.

NOTE 24—One millilitre of standard phosphate solution B/250 mL of solution is equivalent to 0.004 % P_2O_5 for a 0.25-g cement sample. Aliquots of 0, 12.5, 25, 50, 74, 100, and 125 mL are equivalent to P_2O_5 contents in the sample of 0, 0.05, 0.10, 0.20, 0.30, 0.40, and 0.50 %.

9.4.2 Prepare a blank by adding 25.0 mL of the standard HCl to a 250-mL volumetric flask and diluting to 250 mL with water.

9.4.3 Develop colors in the series of phosphate solutions, and in the blank, in accordance with 9.4.6 through 9.4.8.

9.4.4 Plot the net absorbance (absorbance of standard minus that of the blank) values obtained as ordinates and the corresponding P_2O_5 concentrations as abscissas. Draw a smooth curve through the points.

NOTE 25—A suitable paper for plotting the calibration curve is a 10 by 15-in. (254 by 381-mm) linear cross section paper having 20 by 20 divisions to the inch. The percentage of P_2O_5 can then be plotted on the long dimension using five divisions equal to 0.01 % P_2O_5. A scale of one division equal to 0.005 absorbance units is suitable as the ordinate (short dimension of the paper). Scales other than this may be used but under no circumstances should a scale division less than 1/20 in. (1.3 mm) be used for 0.005 units of absorbance or for 0.005 % P_2O_5. A separate calibration curve should be made for each spectrophotometer used, and the calibration curve checked against standard phosphate solution whenever a new batch of ammonium molybdate reagent is used.

9.4.5 Transfer 0.250 g of the sample to a 250-mL beaker and moisten with 10 mL of cold water to prevent lumping. Add 25.0 mL of the standard HCl and digest with the aid of gentle heat and agitation until solution is complete. Filter into a 250-mL volumetric flask and wash the paper and the separated silica thoroughly with hot water. Allow the solution to cool and then dilute with water to 250 mL.

9.4.6 Transfer a 50.0-mL aliquot (Note 26) of the sample solution to a 250-mL beaker, add 5.0 mL of ammonium molybdate solution and 0.1 g of ascorbic acid powder. Mix the contents of the beaker by swirling until the ascorbic acid has dissolved completely. Heat the solution to vigorous boiling and then boil, uncovered, for 1.5 ± 0.5 min. Cool to room temperature and transfer to a 50-mL volumetric flask. Rinse the beaker with one small portion of water and add the rinse water to the flask. Dilute to 50 mL with water.

NOTE 26—The range of the test can be extended by taking a smaller aliquot of the sample solution. In such instances the decrease in the aliquot volume must be made up by the blank solution (9.4.5) to maintain the proper acidity of the final solution. Thus, if a 25-mL aliquot of the sample solution is taken (instead of the usual 50 mL), a 25-mL aliquot of the blank solution should be added before proceeding with the test. The result of the test must then be calculated accordingly.

9.4.7 Measure the absorbance of the solution against water as the reference at 725.0 nm.

9.4.8 Develop on a 50.0-mL aliquot of the blank solution prepared in 9.4.2 in the same manner as was used in 9.4.6 for the sample solution. Measure the absorbance in accordance with 9.4.7 and subtract this absorbance value from that obtained for the sample solution in 9.4.6 in order to obtain the net absorbance for the sample solution.

9.4.9 Using the net absorbance value found in 9.4.8, record the percentage of P_2O_5 in the cement sample as indicated by the calibration curve. Report the percentage of P_2O_5 to the nearest 0.01.

10. Titanium Dioxide (*Reference Method*)

10.1 *Summary of Method*—In this method titanium dioxide (TiO_2) in portland cement is determined colorimetrically using Tiron reagent. Under the conditions of the test iron is the only constituent of portland cement causing a very slight interference equivalent to 0.01 % for each 1 % of Fe_2O_3 present in the sample.

10.2 *Apparatus:*

10.2.1 *Spectrophotometer* (Note 27):

10.2.1.1 The instrument shall be equipped to measure absorbance of solutions at a spectral wavelength of 410 nm.

10.2.1.2 Wavelength measurements shall be repeatable within ±1 nm or less.

10.2.1.3 In the absorbance range from 0.1 to 1.0, the absorbance measurements shall be repeatable within ±1 % or less.

10.2.1.4 To establish that the spectrophotometer will permit a satisfactory degree of accuracy, qualify the instrument in accordance with 3.3.2 using the procedure in 10.4.1 through 10.4.6 of this method.

NOTE 27—For the measurement of the performance of the spectrophotometer, refer to Practice E 275.

10.3 *Reagents:*

10.3.1 *Buffer* (pH 4.7)—68 g of $NaC_2H_3O_2 \cdot 3H_2O$, plus 380 mL of water, plus 100 mL of 5.0 N CH_3COOH.

10.3.2 *Ethylenedinitrilo Tetraacetic Acid Disodium Salt, Dihydrate* (0.2 M EDTA)—Dissolve 37.5 g of EDTA in 350 mL of warm water, and filter. Add 0.25 g of $FeCl_3 \cdot 6H_2O$ and dilute to 500 mL.

10.3.3 *Hydrochloric Acid* (1+6).

10.3.4 *Hydrochloric Acid, Standard* (6.5 N)—Dilute 540 mL of concentrated HCl (sp gr 1.19) to 1 L with water.

10.3.5 *Ammonium Hydroxide* (NH_4OH, 1+1).

10.3.6 *Potassium Pyrosulfate* ($K_2S_2O_7$).

10.3.7 *Titanium Dioxide, Stock Solution A*—Fuse slowly in a platinum crucible over a very small flame 0.0314 g of NBS SRM 154b (TiO_2 = 99.74 %) or its replacements with about 2 or 3 g of $K_2S_2O_7$. Allow to cool, and place the crucible in a beaker containing 125 mL of H_2SO_4 (1+1). Heat and stir until the melt is completely dissolved. Cool, transfer to a 250-mL volumetric flask, and dilute the solution to volume.

10.3.7.1 *Titanium Dioxide, Dilute Standard Solution B* (1 mL = 0.0125 mg TiO_2).—Pipet 50 mL of stock TiO_2 solution into a 500-mL volumetric flask, and dilute to volume. One millilitre of this solution is equal to 0.0125 mg of TiO_2, which is equivalent to 0.05 % TiO_2 when used as outlined in 10.4.4 through 10.4.6.

10.3.8 *Sulfuric Acid* (1+1).

10.3.9 *Tiron* (disodium-1, 2-dihydroxybenzene-3, 5 disulfonate).

10.4 *Procedure:*

10.4.1 Prepare a series of TiO_2 solutions to cover the range from 0 to 1.0 % TiO_2. Prepare each solution in a 50-mL volumetric flask.

NOTE 28—One millilitre of dilute TiO_2 standard solution B per 50 mL (10.3.7.1) is equivalent to 0.05 % TiO_2 for a 0.2500-g cement sample. Aliquots of 0, 5, 10, 15, and 20 mL of dilute TiO_2 standard solution are equivalent to TiO_2 contents in the sample of 0,

0.25, 0.50, 0.75, and 1.0 %. Dilute each to 25 mL with water.

10.4.2 Develop color in accordance with 10.4.4 starting with second sentence. Measure absorbance in accordance with 10.4.5.

10.4.3 Plot absorbance values obtained as ordinates and the corresponding TiO_2 concentrations as abscissas. Draw a smooth curve through the points.

NOTE 29—A suitable paper for plotting the calibration curve is a linear cross section paper having 10 x 10 divisions to 1 cm. A scale division equivalent to 0.002 absorbance and 0.002 % TiO_2 should be used. A separate calibration curve should be made for each spectrophotometer used.

10.4.4 Transfer a 25.0-mL aliquot of the sample solution prepared in 9.4.5 into a 50-mL volumetric flask (Note 30). Add 5 mL tiron and 5 mL EDTA, mix, and then add NH_4OH (1+1) dropwise, mixing thoroughly after each drop, until the color changes through yellow to green, blue, or ruby red. Then, *just restore* the yellow color with HCl (1+6) *added dropwise and mixing after each drop.* Add 5 mL buffer, dilute to volume and mix.

10.4.5 Measure the absorbance of the solution against water as the reference at 410 nm.

NOTE 30—The range of the test can be extended by taking a smaller aliquot. The results of the test must then be calculated accordingly.

10.4.6 Using the absorbance value determined in 10.4.5, record the percentage of TiO_2 in the cement sample as indicated by the calibration curve to the nearest 0.01. Correct for the iron present in the sample to obtain the true TiO_2 as follows: True TiO_2 = measured % TiO_2 − (0.01 × % Fe_2O_3). Report the percent of TiO_2 to the nearest 0.01.

11. Zinc Oxide (*Reference Method*)

11.1 *Summary of Method:*

11.1.1 This method is applicable to the determination of zinc oxide (ZnO) in portland cement. ZnO is separated from an acidic solution using dithizone as the extractant. After purification steps, the color intensity is measured on a spectrophotometer or colorimeter.

11.1.2 **Caution**—This method requires the use of carbon tetrachloride (CCl_4) which is a hazardous substance with carcinogenic potential for man. Consequently, anyone using this method shall be thoroughly familiar with all regulatory, safety, and health requirements and precautions regarding the receipt, storage, use, and disposal of this material (see Note 32).

11.2 *Apparatus:*

11.2.1 *Spectrophotometer (Note 31)*—The instrument shall be equipped to measure absorbance of solutions at a spectral wavelength of 535 nm. Wavelength measurements shall be repeatable to within ±1 nm. In the absorbance range from 0.1 to 1.0, the absorbance measurements shall be repeatable to within ±1 %. To establish that the spectrophotometer will permit a satisfactory degree of accuracy, qualify the instrument in accordance with 3.3.2 using the procedure in 11.4.1 through 11.5.

NOTE 31—For the measurement of the performance of the spectrophotometer, refer to Practice E 275.

11.3 *Reagents:*

11.3.1 *Zinc-Free Water*—Redistill distilled water using a still composed of borosilicate glass, or pass distilled water through a demineralizer.

11.3.2 *Carbon Tetrachloride* (CCl_4)—Use analytical reagent grade CCl_4 without purification. Purification of technical or used CCl_4 is not recommended.

NOTE 32—Carbon tetrachloride is a known carcinogen. Avoid breathing vapors and use extreme care while handling. Alternate solvents are being investigated.

11.3.3 *Dithizone Reagent*—Dissolve 0.2 g of diphenylthiocarbazone (dithizone) in a beaker containing 1 L of CCl_4 and transfer to a 4-L separatory funnel. Agitate the contents frequently for 15 min. Add 2 L of 0.02 N NH_4OH and extract the dithizone into the aqueous phase by vigorous shaking. Discard the CCl_4 phase and extract the aqueous solution of dithizone with 50-mL portions of CCl_4 until the extracts are a pure green color. Discard the CCl_4 from each extraction. Add 500 mL of CCl_4 and 50 mL of 1 N HCl. Shake the funnel vigorously to transfer the dithizone into the CCl_4 phase. Transfer the CCl_4 phase to a 2-L volumetric flask and dilute to volume with CCl_4. Store the solution in a dark, cool place in a low actinic borosilicate glass bottle or in a clear borosilicate glass bottle painted black. This solution is stable for one week.

11.3.4 *Ammonium Citrate Buffer*—Dissolve 225 g of ammonium citrate in a beaker containing 1 L of water. Add concentrated NH_4OH until the solution has a pH of 8.5 (Note 33). Transfer the solution to a 4-L separatory funnel and purify the extraction with two 100-mL portions of dithizone reagent, followed by two extractions with 100 mL of CCl$_4$.

NOTE 33—Proper preparation of the dithizone solution and subsequent extractions of sample solutions (see 11.4.3) are pH dependent. For this reason the NH$_4$OH and HCl should be at the proper normality, and the buffer should be at the proper pH.

11.3.5 *Phenolphthalein Indicator*—0.1 % solution in 90 % ethanol.

11.3.6 *Carbamate Reagent*—Dissolve 2.5 g of sodiumdiethyldithiocarbamate in 1 L of zinc-free water. Shake the solution with a 50-mL portion of CCl$_4$ to remove Cu, discarding the CCl$_4$ phase. This solution may be stored for one month in a refrigerator.

11.3.7 *Standard Zinc Stock Solution* (100 ppm)—Place exactly 0.1000 g of pure zinc metal (30 mesh, analytical reagent) in a 1-L volumetric flask. Add 50 mL of zinc-free water and 1 mL of concentrated H$_2$SO$_4$. When the zinc has dissolved, dilute to volume with zinc-free water.

11.3.8 *Standard Zinc Working Solution* (2 ppm)—Transfer by pipette 10 mL of the zinc stock solution to a 500-mL volumetric flask. Add 25 mL of HCl (1+1) and dilute to volume.

11.3.9 *Ammonium Hydroxide* (NH$_4$OH), 0.01 N and 0.02 N—Prepare using concentrated NH$_4$OH and zinc-free water.

11.3.10 *Hydrochloric Acid* (HCl), 1 N and 0.02 N—Prepare using concentrated HCl and zinc-free water (see Note 33).

11.4 *Procedure:*

11.4.1 Transfer 0.200 g of the sample to a 250-mL beaker. Add about 50 mL of zinc-free water and 10 mL of HCl (1+1). Digest with the aid of gentle heat and agitation until solution is complete. Quantitatively transfer the solution to a 200-mL volumetric flask and dilute to volume with zinc-free water. This solution now contains 1000 ppm of total solids. Prepare a blank by adding 10 mL of HCl (1+1) to a 200-mL volumetric flask and diluting to volume with zinc-free water.

11.4.2 Transfer by pipette a 10 mL aliquot of sample solution, zinc working solution, and a blank into individual 125-mL Squibb-type separatory funnels. Add two drops of phenophthalein indicator and add concentrated NH$_4$OH dropwise until the solution turns faint pink (pH 8.3 to 8.5).

11.4.3 Add 10 mL of dithizone reagent, shake the funnel three minutes on a shaker, allow the phases to separate, and drain off the CCl$_4$ phase into a second separatory funnel containing 50 mL of 0.02 N HCl. Make a second extraction with 10 mL of dithizone reagent, combining it with the first 10 mL portion. An appreciable blue color in the second dithizone extract indicates that insufficient dithizone has been added to extract all of the dithizonate-forming metals from the solution. In such a case, make a third extraction with 10 mL of dithizone reagent. Finally, add 5 mL of CCl$_4$, shake the funnel 10 times by hand, and combine the CCl$_4$ phase with the other extracts in the second separatory funnel. The zinc and other dithizonate-forming metals are now in the CCl$_4$. Discard the ammonical aqueous phase in the first set of separatory funnels.

11.4.4 Stopper the funnel containing the 50 mL of 0.02 N HCl and dithizonates in CCl$_4$ and shake 2 min. Zinc and certain other metals are now in the aqueous phase. Allow the phases to separate and discard the CCl$_4$ phase, which contains Cu and other acid-stable dithizonates. Rinse the aqueous phase once with 5 mL of CCl$_4$, discarding it as before.

11.4.5 To the funnel containing the 50 mL of 0.02 N HCl, add 5 mL of ammonium citrate buffer and two drops of phenophthalein indicator. Titrate the solution to just pink with concentrated NH$_4$OH. Add 10 mL of carbamate reagent followed by 10.0 mL of dithizone reagent. Shake the funnel 4 min, allow the phases to separate, and drain off the CCl$_4$ phase into a third separatory funnel containing 50 mL of the 0.01 N NH$_4$OH. During this extraction, the zinc is extracted quantitatively into the CCl$_4$ phase, whereas all other interfering elements are retained in the aqueous phase.

11.4.6 Shake for 2 min the funnel containing the 50 mL of 0.01 N NH$_4$OH and zinc dithizonate in CCl$_4$. Allow the phases to separate and drain a portion of the violet-red zinc dithiozonate extract into a beaker. Pipette 5.0 mL of this solution into a 25-mL volumetric flask. Make the solution to volume with CCl$_4$, mix it thoroughly, and measure the absorbance of the sample, standard, and blank extracts against CCl$_4$ as the reference at 535 nm. Correct the unknown and the standard for the zinc contained in the reagent blank.

NOTE 34—Protect the final extractions from light as much as possible and read within 2 h.

11.5 *Calculation:*

11.5.1 Calculate the percentage ZnO to the nearest 0.01 % as follows:

$$\%, \text{ZnO} = \frac{\text{Absorbance Sample}}{\text{Absorbance Standard}} \times \frac{2.0}{1000} \times 1.2447$$

$$= \frac{\text{Absorbance Sample}}{\text{Absorbance Standard}} \times 0.25$$

NOTE 35—If the ZnO content of the sample measures higher than 0.25 %, redilute the sample with CCl₄ so that it is below the standard and determine its absorbance again.

12. Aluminum Oxide (*Reference Method*)

NOTE 36—In the reference method, Al_2O_3 is calculated from the ammonium hydroxide group by subtracting the separately determined constituents that usually are present in significant amounts in the ammonium hydroxide precipitate. These are Fe_2O_3, TiO_2 and P_2O_5. Most instrumental methods for Al_2O_3 analysis give Al_2O_3 alone if standardized and calibrated properly.

12.1 *Calculation:*

12.1.1 Calculate the percentage of Al_2O_3 by deducting the percentage of the sum of the Fe_2O_3, TiO_2, and P_2O_5 from the percentage of ammonium hydroxide group. All determinations shall be by referee methods described in the appropriate sections herein. All percentages shall be calculated to the nearest 0.01 %. Report the Al_2O_3 to the nearest 0.1 %. For nonreferee analyses, the percentages of Fe_2O_3, TiO_2, and P_2O_5 can be determined by any procedure for which qualification has been shown.

13. Calcium Oxide (*Reference Method*)

13.1 *Summary of Method:*

13.1.1 In this method, manganese is removed from the filtrate after the determination of SiO_2 and the ammonium hydroxide group. Calcium is then precipitated as the oxalate. After filtering, the oxalate is redissolved and titrated with potassium permanganate ($KMnO_4$).

NOTE 37—For referee analysis or for the most accurate determinations, removal of manganese in accordance with 13.3.2 must be made. For less accurate determinations, and when only insignificant amounts of manganese oxides are believed present, 13.3.2 may be omitted.

13.1.2 Strontium, usually present in portland cement as a minor constituent, is precipitated with calcium as the oxalate and is subsequently titrated and calculated as CaO. If the SrO content is known and correction of CaO for SrO is desired as, for example, for research purposes or to compare results with SRM certificate values, the CaO obtained by this method may be corrected for SrO. In determining conformance of a cement to specifications, the correction of CaO for SrO should not be made.

13.2 *Reagents:*

13.2.1 *Ammonium Oxalate Solution* (50 g/L).

13.2.2 *Potassium Permanganate, Standard Solution* (0.18 N)—Prepare a solution of potassium permanganate ($KMnO_4$) containing 5.69 g/L. Let this solution stand at room temperature for at least 1 week, or boil and cool to room temperature. Siphon off the clear solution without disturbing the sediment on the bottom of the bottle; then filter the siphoned solution through a bed of glass wool in a funnel or through a suitable sintered glass filter. Do not filter through materials containing organic matter. Store in a dark bottle, preferably one that has been painted black on the outside. Standardize the solution against 0.7000 to 0.8000 g of primary standard sodium oxalate, according to the directions furnished with the sodium oxalate and record the temperature at which the standardization was made (Note 38).

13.2.2.1 Calculate the CaO equivalent of the solution as follows:

1 mL of 1 N $KMnO_4$ solution is equivalent to 0.06701 g of pure sodium oxalate.

$$\text{Normality of } KMnO_4 = \frac{\text{weight of sodium oxalate} \times \text{fraction of its purity}}{\text{mL of } KMnO_4 \text{ solution} \times 0.06701}$$

1 mL of 1 N $KMnO_4$ solution is equivalent to 0.02804 g of CaO.

$$F = \frac{\text{normality of } KMnO_4 \text{ solution} \times 0.02804 \times 100}{0.5}$$

where F = CaO equivalent of the $KMnO_4$ solution in % CaO/mL based on a 0.5-g sample of cement.

NOTE 38—Because of the instability of the $KMnO_4$ solution, it is recommended that it be restandardized at least bimonthly.

13.3 *Procedure:*

13.3.1 Acidify the combined filtrates obtained in the precipitations of the ammonium hydroxide group (7.2.2). Neutralize with HCl to the methyl red end point, make just acid, and add 6 drops of HCl in excess.

13.3.2 *Removal of Manganese*—Evaporate to

a volume of about 100 mL. Add 40 mL of saturated bromine water to the hot solution and immediately add NH_4OH until the solution is distinctly alkaline. Addition of 10 mL of NH_4OH is generally sufficient. A piece of filter paper, about 1 cm^2 in area, placed in the heel of the beaker and held down by the end of a stirring rod aids in preventing bumping and initiating precipitation of hydrated manganese oxides (MnO). Boil the solution for 5 min or more, making certain that the solution is distinctly alkaline at all times. Allow the precipitate to settle, filter using medium-textured paper, and wash with hot water. If a precipitate does not appear immediately, allow a settling period of up to 1 h before filtration. Discard any manganese dioxide that may have been precipitated. Acidify the filtrate with HCl using litmus paper as an indicator, and boil until all the bromine is expelled (Note 39).

13.3.3 Add 5 mL of HCl, dilute to 200 mL, and add a few drops of methyl red indicator and 30 mL of warm ammonium oxalate solution (50 g/L). (Note 40). Heat the solution to 70 to 80°C, and add NH_4OH (1+1) dropwise, while stirring until the color changes from red to yellow (Note 41). Allow the solution to stand without further heating for 60 ± 5 min (no longer), with occasional stirring during the first 30 min.

13.3.4 Filter, using retentive paper, and wash the precipitate 8 to 10 times with hot water, the total amount of water used in rinsing the beaker and washing not to exceed 75 mL. During this washing, water from the wash bottle should be directed around the inside of the filter paper to wash the precipitate down, then a jet of water should be gently directed towards the center of the paper in order to agitate and thoroughly wash the precipitate. Acidify the filtrate with HCl and reserve for the determination of MgO.

13.3.5 Place the beaker in which the precipitation was made under the funnel, pierce the apex of the filter paper with the stirring rod, place the rod in the beaker, and wash the precipitate into the beaker by using a jet of hot water. Drop about 10 drops of H_2SO_4 (1+1) around the top edge of the filter paper. Wash the paper five more times with hot water. Dilute to 200 mL, and add 10 mL of H_2SO_4 (1+1). Heat the solution to a temperature just below boiling, and titrate it immediately with the 0.18 N $KMnO_4$ solution (Note 42). Continue the titration slowly until the pink color persists for at least 10 s. Add the filter paper that contained the original precipitate and macerate it. If the pink color disappears continue the titration until it again persists for at least 10 s.

NOTE 39—Potassium iodide starch paper may be used to indicate the complete volatilization of the excess bromine. Expose a strip of moistened paper to the fumes from the boiling solution. The paper should remain colorless. If it turns blue bromine is still present.

NOTE 40—If the ammonium oxalate solution is not perfectly clear, it should be filtered before use.

NOTE 41—This neutralization must be made slowly, otherwise precipitated calcium oxalate may have a tendency to run through the filter paper. When a number of these determinations are being made simultaneously, the following technique will assist in ensuring slow neutralization. Add two or three drops of NH_4OH to the first beaker while stirring, then 2 or 3 drops to the second, and so on, returning to the first beaker to add 2 or 3 more drops, etc., until the indicator color has changed in each beaker.

NOTE 42—The temperature of the 0.18 N $KMnO_4$ solution at time of use should not vary from its standardization temperature by more than 10°F (5.5°C). Larger deviations could cause serious error in the determination of CaO.

13.3.6 *Blank*—Make a blank determination, following the same procedure and using the same amounts of reagents (Note 43), and record the millilitres of $KMnO_4$ solution required to establish the end point.

NOTE 43—When the amount of calcium oxalate is very small, its oxidation by $KMnO_4$ is slow to start. Before the titration, add a little $MnSO_4$ to the solution to catalyze the reaction.

13.4 *Calculation:*

13.4.1 Calculate the percentage of CaO to the nearest 0.1 as follows:

$$CaO, \% = E(V - B)$$

where:

E = CaO equivalent of the $KMnO_4$ solution in % CaO/mL based on a 0.5-g sample,

V = millilitres of $KMnO_4$ solution required by the sample, and

B = millititres of $KMnO_4$ solution required by the blank.

13.4.2 If desired calculate the percentage of CaO corrected for SrO as follows:

$$CaO_c \% = CaO_i \% - 0.54 \, SrO \%$$

where:

CaO_c = CaO corrected for SrO, and
CaO_i = initial CaO as determined in 13.4.1

$$0.54 = \frac{56.08}{103.62} \frac{CaO}{SrO} = \text{molecular weight ratio}$$

14. Magnesium Oxide (*Reference Method*)

14.1 *Summary of Method*—In this method, magnesium is precipitated as magnesium ammonium phosphate from the filtrate after removal of calcium. The precipitate is ignited and weighed as magnesium pyrophosphate ($Mg_2P_2O_7$). The MgO equivalent is then calculated.

14.2 *Reagent*—Ammonium phosphate, dibasic (100 g/L) $(NH_4)_2HPO_4$.

14.3 *Procedure:*

14.3.1 Acidify the filtrate from the determination of CaO (13.3.4) with HCl and evaporate by boiling to about 250 mL. Cool the solution to room temperature, add 10 mL of ammonium phosphate, dibasic, $(NH_4)_2HPO_4$ (100 g/L), and 30 mL of NH_4OH. Stir the solution vigorously during the addition of NH_4OH and then for 10 to 15 min longer. Let the solution stand for at least 8 h in a cool atmosphere and filter. Wash the residue five or six times with NH_4OH (1+20) and ignite in a weighed platinum or porcelain crucible, at first slowly until the filter paper is charred and then burn off (see Note 44), and finally at 1100°C for 30 to 45 min. Weigh the residue as magnesium pyrophosphate ($Mg_2P_2O_7$).

14.3.2 *Blank*—Make a blank determination following the same procedure and using the same amounts of reagents, and correct the results obtained in the analysis accordingly.

14.4 *Calculation:*

14.4.1 Calculate the percentage of MgO to the nearest 0.1 as follows:

$$MgO, \% = W \times 72.4$$

where:

W = grams of $Mg_2P_2O_7$, and
72.4 = molecular ratio of 2MgO to $Mg_2P_2O_7$ (0.362) divided by the weight of sample used (0.5 g) and multiplied by 100.

NOTE 44—Extreme caution should be exercised during this ignition. Reduction of the phosphate precipitate can result if carbon is in contact with it at high temperatures. There is also danger of occluding carbon in the precipitate if ignition is too rapid.

15. Sulfur (See Note 45)

15.1 *Sulfur Trioxide:* (*Reference Method*)

15.1.1 *Summary of Method*—In this method, sulfate is precipitated from an acid solution of the cement with barium chloride ($BaCl_2$). The precipitate is ignited and weighed as barium sulfate ($BaSO_4$) and the SO_3 equivalent is calculated.

15.1.2 *Procedure:*

15.1.2.1 To 1 g of the sample add 25 mL of cold water and, while the mixture is stirred vigorously, add 5 mL of HCl (Note 46). If necessary, heat the solution and grind the material with the flattened end of a glass rod until it is evident that decomposition of the cement is complete (Note 47). Dilute the solution to 50 mL and digest for 15 min at a temperature just below boiling. Filter through a medium-textured paper and wash the residue thoroughly with hot water. Dilute the filtrate to 250 mL and heat to boiling. Add slowly, dropwise, 10 mL of hot $BaCl_2$ (100 g/L) and continue the boiling until the precipitate is well formed. Digest the solution for 12 to 24 h at a temperature just below boiling (Note 48). Take care to keep the volume of solution between 225 and 260 mL and add water for this purpose if necessary. Filter through a retentive paper, wash the precipitate thoroughly with hot water, place the paper and contents in a weighed platinum crucible, and slowly char and consume the paper without inflaming. Ignite at 800 to 900°C, cool in a desiccator, and weigh.

NOTE 45—When an instrumental method is used for sulfur or when comparing results of classical wet and instrumental methods, consult 4.1.2 of these methods.

NOTE 46—The acid filtrate obtained in the determination of the insoluble residue (5.3.1) may be used for the determination of SO_3 instead of using a separate sample.

NOTE 47—A brown residue due to compounds of manganese may be disregarded (see Note 11).

NOTE 48—If a rapid determination is desired, the time of digestion may be reduced to as little as 3 h. However, the cement may be rejected for failure to meet the specification requirement only on the basis of results obtained when using 12 to 24-h digestion times.

15.1.2.2 *Blank*—Make a blank determination following the same procedure and using the same amounts of reagents, and correct the results obtained in the analysis accordingly.

15.1.3 *Calculation*—Calculate the percentage of SO_3 to the nearest 0.01 as follows:

$$SO_3, \% = W \times 34.3$$

where:

W = grams of $BaSO_4$, and

34.3 = molecular ratio of SO_3 to $BaSO_4$ (0.343) multiplied by 100.

15.2 Sulfide: (Reference Method)

15.2.1 Summary of Method
—In this method sulfide sulfur is determined by evolution as hydrogen sulfide (H_2S) from an acid solution of the cement into a solution of ammoniacal zinc sulfate ($ZnSO_4$) or cadmium chloride ($CdCl_2$). The sulfide sulfur is then titrated with a standard solution of potassium iodate (KIO_3). Sulfites, thiosulfates, and other compounds intermediate between sulfides and sulfates are assumed to be absent. If such compounds are present, they may cause an error in the determination.

15.2.2 Apparatus:

15.2.2.1 *Gas-Generating Flask*—Connect a dry 500-mL boiling flask with a long-stem separatory funnel and a small connecting bulb by means of a rubber stopper. Bend the stem of the funnel so that it will not interfere with the connecting bulb, adjust the stem so that the lower end is close to the bottom of the flask, and connect the opening of the funnel with a source of compressed air. Connect the bulb with an L-shaped glass tube and a straight glass tube about 200 mm in length. Insert the straight glass tube in a tall-form, 400-mL beaker. A three-neck distilling flask with a long glass tubing in the middle opening, placed between the source of compressed air and the funnel, is a convenient aid in the regulation of the airflow. Rubber used in the apparatus shall be pure gum grade, low in sulfur, and shall be cleaned with warm HCl.

15.2.3 Reagents:

15.2.3.1 *Ammoniacal Cadmium Chloride Solution*—Dissolve 15 g of cadmium chloride ($CdCl_2 \cdot 2H_2O$) in 150 mL of water and 350 mL of NH_4OH. Filter the solution after allowing it to stand at least 24 h.

15.2.3.2 *Ammoniacal Zinc Sulfate Solution*—Dissolve 50 g of zinc sulfate ($ZnSO_4 \cdot 7H_2O$) in 150 mL of water and 350 mL of NH_4OH. Filter the solution after allowing it to stand at least 24 h.

15.2.3.3 *Potassium Iodate, Standard Solution* (0.03 N)—Prepare a solution of potassium iodate (KIO_3) and potassium iodide (KI) as follows: Dry KIO_3 at 180°C to constant weight. Weigh 1.0701 g of the KIO_3 and 12 g of KI. Dissolve and dilute to 1 L in a volumetric flask. This is a primary standard and requires no standardization (Note 49). One millilitre of this solution is equivalent to 0.0004809 g of sulfur.

NOTE 49—The solution is very stable, but may not maintain its titer indefinitely. Whenever such a solution is over 1 year old it should be discarded or its concentration checked by standardization.

15.2.3.4 *Stannous Chloride Solution*—To 10 g of stannous chloride ($SnCl_2 \cdot 2H_2O$) in a small flask, add 7 mL of HCl (1 + 1), warm the mixture gently until the salt is dissolved, cool the solution, and add 95 mL of water. This solution should be prepared as needed, as the salt tends to hydrolyze.

15.2.3.5 *Starch Solution*—To 100 mL of boiling water, add a cool suspension of 1 g of soluble starch in 5 mL of water and cool. Add a cool solution of 1 g of sodium hydroxide (NaOH) in 10 mL of water, then 3 g of potassium iodide (KI), and mix thoroughly.

15.2.4 Procedure:

15.2.4.1 Place 15 mL of the ammoniacal $ZnSO_4$ or $CdCl_2$ solution (Note 50) and 285 mL of water in a beaker. Put 5 g of the sample (Note 51) and 10 mL of water in the flask and shake the flask gently to wet and disperse the cement completely. This step and the addition of $SnCl_2$ should be performed rapidly to prevent the setting of the cement. Connect the flask with the funnel and bulb. Add 25 mL of the $SnCl_2$ solution through the funnel and shake the flask. Add 100 mL of HCl (1 + 3) through the funnel and shake the flask. During these shakings keep the funnel closed and the delivery tube in the ammoniacal $ZnSO_4$ or $CdCl_2$ solution. Connect the funnel with the source of compressed air, open the funnel, start a slow stream of air, and heat the flask and contents slowly to boiling. Continue the boiling gently for 5 or 6 min. Cut off the heat, and continue the passage of air for 3 or 4 min. Disconnect the delivery tube and leave it in the solution for use as a stirrer. Cool the solution to 20 to 30°C (Note 52), add 2 mL of the starch solution and 40 mL of HCl (1 + 1) and titrate immediately with the 0.03 N KIO_3 solution until a persistent blue color is obtained (Note 53).

NOTE 50—In general, the $ZnSO_4$ is preferable to the $CdCl_2$ solution because $ZnSO_4$ is more soluble in NH_2OH than is $CdCl_2$. The $CdCl_2$ solution may be used when there is doubt as to the presence of a trace of sulfide sulfur, as the yellow cadmium sulfide (CdS) facilitates the detection of a trace.

NOTE 51—If the content of sulfur exceeds 0.20 or 0.25 %, a smaller sample should be used so that the titration with the KIO_3 solution will not exceed 25 mL.

NOTE 52—The cooling is important as the end point

is indistinct in a warm solution.

NOTE 53—If the content of sulfur is appreciable but not approximately known in advance, the result may be low due to the loss of H_2S during a slow titration. In such a case the determination should be repeated with the titration carried out more rapidly.

15.2.4.2 Make a blank determination, following the same procedure and using the same amounts of reagents. Record the volume of KIO_3 solution necessary to establish the end point as described in 15.2.4.1.

15.2.5 *Calculation*—Calculate the percentage of sulfide sulfur (see 15.2.1) as follows:

$$\text{Sulfide}, \% = E(V - B) \times 20$$

where:
E = sulfide equivalent of the KIO_3 solution, g/mL,
V = millilitres of KIO_3 solution required by the sample,
B = millilitres of KIO_3 solution required by the blank, and
20 = 100 divided by the weight of sample used (5 g).

16. Loss on Ignition (*Reference Methods*)

16.1 *Portland Cement:*

16.1.1 *Summary of Method*—In this method, the cement is ignited in a muffle furnace at a controlled temperature. The loss is assumed to represent the total moisture and CO_2 in the cement. This procedure is not suitable for the determination of the loss on ignition of portland blast-furnance slag cement and of slag cement. A method suitable for such cements is described in 16.2.1 through 16.2.3.

16.1.2 *Procedure*—Weigh 1 g of the sample in a tared platinum crucible. Cover and ignite the crucible and its contents to constant weight in a muffle furnace at a temperature of 950 ± 50°C. Allow a minimum of 15 min for the initial heating period and at least 5 min for all subsequent periods.

16.1.3 *Calculation*—Calculate the percentage of loss on ignition to the nearest 0.1 by multiplying the loss of weight in grams by 100.

16.2 *Portland Blast-Furnace Slag Cement and Slag Cement:*

16.2.1 *Summary of Method*—Since it is desired that the reported loss on ignition represent moisture and CO_2, this method provides a correction for the gain in weight due to oxidation of sulfides usually present in portland blast-furnace slag cement and slag cement by determining the increase in SO_3 content during ignition. An optional method providing for a correction based on the decrease in sulfide sulfur during ignition is given in 23.1.1 through 23.1.3.1.

16.2.2 *Procedure:*

16.2.2.1 Weigh 1 g of cement into a tared platinum crucible and ignite in a muffle furnace at a temperature of 950 ± 50°C for 15 min. Cool to room temperature in a desiccator and weigh. Without checking for constant weight, carefully transfer the ignited material to a 400-mL beaker. Break up any lumps in the ignited cement with the flattened end of a glass rod.

16.2.2.2 Determine the SO_3 content by the method given in 15.1.1.1 through 15.1.3.1 (Note 54). Also determine the SO_3 content of a portion of the same cement that has not been ignited, using the same procedure.

NOTE 54—Some of the acid used for dissolving the sample may first be warmed in the platinum crucible to dissolve any adhering material.

16.2.3 *Calculation*—Calculate the percentage loss of weight occurring during ignition and add 0.8 times the difference between the percentages of SO_3 in the ignited sample and the original cement (Note 55). Report the corrected percentage as loss on ignition.

NOTE 55—If a gain in weight is obtained during ignition, subtract the percentage gain from the correction for SO_3.

17. Sodium and Potassium Oxides (*Reference Methods*)

17.1 *Total Alkalies:*

17.1.1 *Summary of Method*—This method[9] covers the determination of sodium oxide (Na_2O) and potassium oxide (K_2O) by flame photometry or atomic absorption.

NOTE 56—This method is suitable for hydraulic cements that are completely decomposed by hydrochloric acid and should not be used for determination of total alkalies in hydraulic cements that contain large amounts of acid-insoluble material, for example, pozzolan cements. It may be used to determine acid-soluble alkalies for such cements. An alternate method of sample dissolution for such cements is in preparation.

---

[9] The 1963 revision of these methods deleted the classical (J. L. Smith) gravimetric method for the determination of Na_2O and K_2O in cements. Those interested in this method should refer to the *1961 Book of ASTM Standards*, Part 4.

The 1983 revision of these methods deleted the details of the flame photometric procedure for the determination of Na_2O and K_2O. Those interested in this method should refer to the *1982 Annual Book of ASTM Standards*, Part 13.

17.1.2 *Apparatus:*

17.1.2.1 *Instrument*—Any type flame photometer or atomic absorption unit may be used provided it can be demonstrated that the required degree of accuracy and precision is as indicated in 17.1.3.

NOTE 57—After such accuracy is established, for a specific instrument, further tests of instrument accuracy are not required except when it must be demonstrated that the instrument gives results within the prescribed degree of accuracy by a single series of tests using the designated standard samples.

NOTE 58—For normal laboratory testing, it is recommended that the accuracy of the instrument be routinely checked by the use of either a National Bureau of Standards cement or cement of known alkali content.

17.1.2.2 The instrument shall consist at least of an atomizer and burner; suitable pressure-regulating devices and gages for fuel and oxidant gas; an optical system, capable of preventing excessive interference from wavelengths of light other than that being measured; and a photosensitive indicating device.

17.1.3 *Initial Qualification of Instrument*—Qualify the instrument in accordance with 3.3.2 to establish that an instrument provides the desired degree of precision and accuracy.

17.1.4 *Reagents and Materials:*

17.1.4.1 *Laboratory Containers*—All glassware shall be made of borosilicate glass and all polyethylene shall comply with the requirements of 4.2.3.

17.1.4.2 *Calcium Carbonate*—The calcium carbonate ($CaCO_3$) used in the preparation of the calcium chloride stock solution (17.1.5.1) shall contain not more than 0.020 % total alkalies as sulfate.

NOTE 59—Materials sold as a primary standard or ACS "low alkali" grade normally meet this requirement. However, the purchaser should assure himself that the actual material used conforms with this requirement.

17.1.4.3 *Potassium Chloride* (KCl).

17.1.4.4 *Sodium Chloride* (NaCl).

17.1.4.5 Commercially available solutions may be used in place of those specified in 17.1.5.

17.1.5 *Preparation of Solutions:*

17.1.5.1 *Calcium Chloride Stock Solution*—Add 300 mL of water to 112.5 g of $CaCO_3$ in a 1500-mL beaker. While stirring, slowly add 500 mL of HCl. Cool the solution to room temperature, filter into a 1-L volumetric flask, dilute to 1 L, and mix thoroughly. This solution contains the equivalent of 63 000 ppm (6.30 %) CaO.

17.1.5.2 *Sodium-Potassium Chloride Stock Solution*—Dissolve 1.8858 g of sodium chloride (NaCl) and 1.583 g of potassium chloride (KCl) (both dried at 105 to 110°C for several hours prior to weighing) in water. Dilute to 1 L in a volumetric flask and mix thoroughly. This solution contains the equivalent of 1000 ppm (0.10 %) each of Na_2O and K_2O. Separate solutions of Na_2O and of K_2O may be used provided that the same concentration solutions are used for calibration for cement analysis as were used for the calibration when qualifying the instrument in accordance with 17.1.3.

17.1.5.3 *Standard Solutions*—Prepare the standard solutions prescribed for the instrument and method used. Measure the required volume of NaCl-KCl stock solutions in calibrated pipets or burets. The calcium chloride stock solutions, if needed, may be measured in appropriate graduated cylinders. If the instrument being used requires an internal standard, measure the internal standard solution with a pipet or buret. Place each solution in a volumetric flask, dilute to the indicated volume, and mix thoroughly.

17.1.5.4 If more dilute solutions are required by the method in use, pipet the required aliquot to the proper sized volumetric flask, add any necessary internal standard, dilute to the mark, and mix thoroughly.

17.1.6 *Calibration of Apparatus:*

NOTE 60—No attempt is made in this section to describe in detail the steps for putting the instrument into operation since this will vary considerably with different instruments. The manufacturer's instructions should be consulted for special techniques or precautions to be employed in the operation, maintenance, or cleaning of the apparatus.

17.1.6.1 Turn on the instrument and allow it to warm up in accordance with the manufacturer's instructions. (A minimum of 30 min is required for most instruments.) Adjust the fuel and oxidant gas pressures as required by the instrument being used. Light and adjust the burner for optimum operation. Make any other adjustments that may be necessary to establish the proper operating conditions for the instrument.

17.1.7 *Procedure:*

17.1.7.1 *Solution of the Cement*—Prepare the solution of the cement in accordance with the procedure specified by the instrument manufacturer. If no procedure is specified, or if desired, proceed as specified in 17.1.7.1.1 or 17.1.7.1.2 (Note 60).

NOTE 61—The presence of SiO_2 in solution affects the accuracy of some flame photometers. In cases where an instrument fails to provide results within the prescribed degree of accuracy outlined in 3.3.2.1 through 3.3.3 tests should be made on solutions from which the SiO_2 has been removed. For this removal proceed as in 17.1.7.1.2.

17.1.7.1.1 Place 1.000 ± 0.001 g of the cement in a 150-mL beaker and disperse with 20 mL of water using a swirling motion of the beaker. While still swirling add 5.0 mL of HCl all at once. Dilute immediately to 50 mL with water. Break up any lumps of cement remaining undispersed with a flat-end stirring rod. Digest on a steam bath or hot plate for 15 min, then filter through a medium-textured filter paper into a 100-mL volumetric flask. Wash beaker and paper thoroughly with hot-water, cool contents of the flask to room temperature, dilute to 100 mL, and mix the solution thoroughly. Continue as given in 17.1.7.2.

17.1.7.1.2 Place 1.000 ± 0.001 g of cement into a platinum evaporating dish and disperse with 10 mL of water using a swirling motion. While still swirling, add 5.0 mL of HCl all at once. Break up any lumps with a flat-end stirring rod and evaporate to dryness on a steam bath. Make certain that the gelatinous appearance is no longer evident. Treat the residue with 2.5 mL of HCl and about 20 mL of water. Digest on a steam bath for 5 to 10 min and filter immediately through a 9-cm medium-textured filter paper into a 100-mL volumetric flask. Wash thoroughly with repeated small amounts of hot water until the total volume of solution is 80 to 95 mL. Cool to room temperature, dilute to the mark, and mix thoroughly.

When it has been demonstrated that the removal of SiO_2 is necessary to obtain the required accuracy described in 3.3.2.1 through 3.3.3 for a specific flame photometer, SiO_2 must always be removed when making analyses that are used as the basis for rejection of a cement for failure to comply with specifications or where specification compliance may be in question. Where there is no question as to specification compliance, analyses may be made by such instruments without SiO_2 removal provided the deviations from certificate values obtained by the tests prescribed in 3.3.2.1 through 3.3.3 are not more than twice the indicated limits.

17.1.7.2 If the method in use requires more dilute solutions, an internal standard, or both, carry out the same dilutions as in 17.1.5.4 as needed. The standard and the sample solutions to be analyzed must be prepared in the same way and to the same dilution as the solutions of standard cements analyzed for the qualification of the instrument.

17.1.7.3 *Procedure for Na_2O* (Note 63)—Warm up and adjust the instrument for the determination of Na_2O as described in 17.1.6.1. Immediately following the adjustment and without changing any instrumental settings, atomize the cement solution and note the scale reading (Note 62). Select the standard solutions which immediately bracket the cement solution in Na_2O content and observe their readings. Their values should agree with the values previously established during calibration of the apparatus. If not, recalibrate the apparatus for that constituent. Finally, alternate the use of the unknown solution and the bracketing standard solutions until readings of the unknown agree within one division on the transmission or meter scale, or within 0.01 weight percent for instruments with digital readout, and readings for the standards similarly agree with the calibration values. Record the average of the last two readings obtained for the unknown solution.

NOTE 62—The order in determining Na_2O or K_2O is optional. In all cases, however, the determination should immediately follow the adjustment of the instrument for that particular constituent.

17.1.7.4 If the reading exceeds the scale maximum, either transfer a 50-mL aliquot of the solution prepared in 17.1.7.1 to a 100-mL volumetric flask or, if desired, prepare a new solution by using 0.500 g of cement and 2.5 mL of HCl (instead of 5.0 mL) in the initial addition of acid. In the event silica has to be removed from the 0.5-g sample of cement, treat the dehydrated material with 1.25 mL of HCl and about 20 mL of water, then digest, filter, and wash. In either case, add 5.0 mL of calcium chloride stock solution (17.1.5.1) before diluting to mark with water. Dilute to the mark. Proceed as in 17.1.5.4 if more dilute solutions are required by the method in use. Determine the alkali content of this solution as described in (17.1.7.3) and multiply by a factor of 2 the percentage of alkali oxide.

17.1.7.5 *Procedure for K_2O*—Repeat the procedure described in 17.1.7.3 except that the instrument shall be adjusted for the determination of K_2O. For instruments that read both Na_2O and K_2O simultaneously, determine K_2O at the same time as determining Na_2O.

17.1.8 *Calculation and Report*—From the recorded averages for Na_2O and K_2O in the unknown sample, report each oxide to the nearest 0.01 %.

17.2 *Water-Soluble Alkalies:*

NOTE 63—The determination of water-soluble alkali should not be considered as a substitute for the determination of total alkali according to 17.1.2.1 to 17.1.8. Moreover, it is not to be assumed that in this method all water-soluble alkali in the cement will be dissolved. Strict adherence to the procedure described is essential where there is a specified limit on the content of water-soluble alkali or where several lots of cement are compared on the basis of water-soluble alkali.

17.2.1 *Procedure:*

17.2.1.1 Weigh 25.0 g of sample into a 500-mL Erlenmeyer flask and add 250 mL of water. Stopper the flask with a rubber stopper and shake continuously for 10 min at room temperature. Filter through a Büchner funnel which contains a well-seated retentive, dry filter paper, into a 500-mL filtering flask, using a weak vacuum. Do not wash.

17.2.1.2 Transfer a 50-mL aliquot (Note 64) of the filtrate to a 100-mL volumetric flask and acidify with 0.5 mL of concentrated HCl (sp gr 1.19). Add 9.0 mL of stock $CaCl_2$ solution (63 000 ppm CaO), described in 17.1.5.1, to the 100-mL flask, and dilute the solution to 100 mL. If the method in use requires more dilute solutions, an internal standard, or both, carry out the same dilutions as in 17.1.5.4, as needed. Determine the Na_2O and K_2O contents of this solution as described in 17.1.7.3 and 17.1.7.5. Record the parts per million of each alkali in the solution in the 100-mL flask.

NOTE 64—The aliquot of the filtrate taken for the analysis should be based on the expected water-soluble alkali content. If the expected level of either K_2O or Na_2O is more than 0.08 weight % of cement, or if the water soluble alkali level is unknown, a 50-mL aliquot as given in 17.2.1.2 should be used to make up the initial test solution. If either the Na_2O or K_2O exceeds 0.16 %, place a 50-mL aliquot of the solution from 17.2.1.2 in a 100-mL volumetric flask, add 5 mL of $CaCl_2$ stock solution, and dilute to 100 mL. When the level of either K_2O or Na_2O is less than 0.08 %, take a 100-mL aliquot from the original filtrate (obtained by 17.2.1.1), add 1 mL of HCl, and evaporate on a hot plate in a 250-mL beaker to about 70 mL. Add 8 mL of stock $CaCl_2$ solution and transfer the sample to a 100-mL volumetric flask, rinsing the beaker with a small portion of distilled water. Cool the solution to room temperature and dilute to 100 mL.

17.2.2 *Calculations*—Calculate the percentage of the water-soluble alkali, expressed as Na_2O, to the nearest 0.01 as follows:

Total water-soluble alkali, as $Na_2O = A + E$

$$A = B/(V \times 10)$$
$$C = D/(V \times 10)$$
$$E = C \times 0.658$$

where:

A = percentage of water-soluble sodium oxide (Na_2O),
V = millilitres of original filtrate in the 100-mL flask,
B = parts per million of Na_2O in the solution in the 100-mL flask,
C = percent of water-soluble potassium oxide (K_2O),
D = parts per million of K_2O in the 100-mL flask,
E = percentage Na_2O equivalent to K_2O determined, and
0.658 = molecular ratio of Na_2O to K_2O.

18. Manganic Oxide (*Reference Method*)

18.1 *Summary of Method*—In this procedure, manganic oxide is determined volumetrically by titration with sodium arsenite solution after oxidizing the manganese in the cement with sodium metabismuthate $(NaBiO_3)$.

18.2 *Reagents:*

18.2.1 *Sodium Arsenite, Standard Solution* (1 mL = 0.0003 g Mn_2O_3)—Dissolve in 100 mL of water 3.0 g of sodium carbonate (Na_2CO_3) and then 0.90 g of arsenic trioxide (As_2O_3), heating the mixture until the solution is as complete as possible. If the solution is not clear or contains a residue, filter the solution. Cool it to room temperature, transfer to a volumetric flask, and dilute to 1 L.

18.2.1.1 Dissolve 0.58 g of potassium permanganate $(KMnO_4)$ in 1 L of water and standardize it against about 0.03 g of sodium oxalate $(Na_2C_2O_4)$ oxidimetric standard furnished by the National Bureau of Standards (Standard Sample No. 40 or its replacement) according to the directions furnished with the sodium oxalate. Put 30.0 mL of the $KMnO_4$ solution in a 250-mL Erlenmeyer flask. Add 60 mL of HNO_3 (1+4) and 10 mL of sodium nitrite $(NaNO_2, 50 g/L)$ to the flask. Boil the solution until the HNO_2 is completely expelled. Cool the solution, add NaBiO$_3$, and finish by titrating with the standard sodium arsenite (NaAsO$_2$) solution as described in 18.3.2. Calculate the manganic oxide (Mn$_2$O$_3$) equivalent of the NaAsO$_2$ solution, g/mL, as follows:

$$E = (A \times 7.08)/BC$$

where:
E = Mn$_2$O$_3$ equivalent of the NaAsO$_2$ solution, g/mL,
A = grams of Na$_2$C$_2$O$_4$ used,
B = millilitres of KMnO$_4$ solution required by the Na$_2$C$_2$O$_4$,
C = millilitres of NaAsO$_2$ solution required by 30.0 mL of KMnO$_4$ solution, and
7.08 = molecular ratio of Mn$_2$O$_3$ to 5 Na$_2$C$_2$O$_4$ (0.236) multiplied by 30.0 (millilitres of KMnO$_4$ solution).

18.2.2 *Sodium Metabismuthate* (NaBiO$_3$).

18.2.3 *Sodium Nitrite Solution* (50 g NaNO$_2$/L).

18.3 *Procedure:*

18.3.1 Weigh 1.0 to 3.0 g of the sample (Note 65) into a 250-mL beaker and treat it with 5 to 10 mL of water and then with 60 to 75 mL of HNO$_3$ (1+4). Boil the mixture until the solution is as complete as possible. Add 10 mL of NaNO$_2$ solution (50 g/L) to the solution and boil it until the nitrous acid is completely expelled (Note 66), taking care not to allow the volume of the solution to become so small as to cause the precipitation of gelatinous SiO$_2$. There may be some separated SiO$_2$, which may be ignored, but if there is still a red or brown residue, use more NaNO$_2$ solution (50 g/L) to effect a complete decomposition, and then boil again to expel the nitrous acid. Filter the solution through a medium-textured paper into a 250-mL Erlenmeyer flask and wash the filter paper with water.

NOTE 65—The amount of cement taken for analysis depends on the content of manganese, varying from 1 g for about 1 % of Mn$_2$O$_3$ to 3 g for 0.25 % or less of Mn$_2$O$_3$.

NOTE 66—When NaNO$_2$ is added, the expulsion of HNO$_2$ by boiling must be complete. If any HNO$_2$ remains in the solution, it will react with the added NaBiO$_3$ and decrease its oxidizing value. If there is any manganese in the cement, the first small quantity of NaBiO$_3$ should bring out a purple color.

18.3.2 The solution should have a volume of 100 to 125 mL. Cool it to room temperature. To the solution add a total of 0.5 g of NaBiO$_3$ in small quantities, while shaking intermittently. After the addition is completed, shake the solution occasionally for 5 min and then add to it 50 mL of cool HNO$_3$ (1+33) which has been previously boiled to expel nitrous acid. Filter the solution through a pad of ignited asbestos in a Gooch crucible or a carbon or fritted-glass filter with the aid of suction. Wash the residue four times with the cool HNO$_3$ (1+33). Titrate the filtrate immediately with the standard solution of NaAsO$_2$. The end point is reached when a yellow color is obtained free of brown or purple tints and does not change upon further addition of NaAsO$_2$ solution.

18.3.3 *Blank*—Make a blank determination, following the same procedure and using the same amounts of reagents, and correct the results obtained in the analysis accordingly.

18.4 Calculate the percentage of Mn$_2$O$_3$ to the nearest 0.01 as follows:

$$\text{Mn}_2\text{O}_3, \% = (EV/S) \times 100$$

where:
E = Mn$_2$O$_3$ equivalent of the NaAsO$_2$ solution, g/mL,
V = millilitres of NaAsO$_2$ solution required by the sample, and
S = grams of sample used.

19. Chloride (*Reference Method*)

19.1 *Summary of Method*—In this method total chloride content of portland cement is determined by the potentiometric titration of chloride with silver nitrate. The procedure is also applicable to hardened concrete, clinker, and portland cement raw mix. Under the conditions of the test, no constituent normally present in these materials will interfere.

NOTE 67—Species that form insoluble silver salts or stable silver complexes in acid solution interfere with potentiometric measurements. Thus, iodides and bromides interfere while fluorides will not. Sulfide salts in concentrations typical of these materials should not interfere because they are decomposed by acid treatment.

19.2 *Apparatus:*

19.2.1 *Chloride, Silver/Sulfide Ion Selective Electrode*, or a silver billet electrode coated with silver chloride (Note 68), with an appropriate reference electrode.

19.2.2 *Potentiometer* with millivolt scale readable to 1 mV or better. A digital read-out is preferred but not required.

19.2.3 *Buret*, Class A, 10-mL capacity with

0.05-mL divisions. A buret of the potentiometric type, having a displaced delivery tip, is convenient, but not required.

NOTE 68—Suitable electrodes are available from Orion, Beckman Instruments, and Leeds and Northrup. Carefully following the manufacturer's instructions, add filling solution to the electrodes. The silver billet electrodes must be coated electrolytically with a thin, even layer of silver chloride. To coat the electrode, dip the clean silver billet of the electrode into a saturated solution of potassium chloride (about 40 g/L) in water and pass an electric current through the electrode from a 1½ to 6-V dry cell with the silver billet electrode connected to the positive terminal of the battery. A carbon rod from an all-dry cell or other suitable electrode is connected to the negative terminal and immersed in the solution to complete the electrical circuit. When the silver chloride coating wears off, it is necessary to rejuvenate the electrode by repeating the above procedure. All of the old silver chloride should first be removed from the silver billet by rubbing it gently with fine emery paper followed by water rinsing of the billet.

19.3 *Reagents:*

19.3.1 *Sodium Chloride* (NaCl), primary standard grade.

19.3.2 *Silver Nitrate* ($AgNO_3$), reagent grade.

19.3.3 *Potassium Chloride* (KCl), reagent grade (required for silver billet electrode only).

19.3.4 *Reagent Water* conforming to the requirements of Specification D 1193 for Type III reagent water.

19.4 *Preparation of Solutions:*

19.4.1 *Sodium Chloride, Standard Solution* (0.05 N NaCl)—Dry sodium chloride (NaCl) at 105 to 110°C to a constant weight. Weigh 2.9222 g of dried reagent. Dissolve in water and dilute to exactly 1 L in a volumetric flask and mix thoroughly. This solution is the standard and requires no further standardization.

19.4.2 *Silver Nitrate, Standard Solution* (0.05 N $AgNO_3$)—Dissolve 8.4938 g of silver nitrate ($AgNO_3$) in water. Dilute to 1 L in a volumetric flask and mix thoroughly. Standardize against 5.00 mL of standard 0.05 N sodium chloride solution diluted to 150 mL with water following the titration method given in 19.5.4 beginning with the second sentence. The exact normality shall be calculated from the average of three determinations as follows:

$$N = 0.25/V$$

where:
N = normality of $AgNO_3$ solution,
0.25 = milliequivalents NaCl (5.00 mL × 0.05 N), and
V = volume of $AgNO_3$ solution, mL.

Commercially available standard solutions may be used provided the normality is checked according to the standardization procedure.

19.4.3 *Methyl Orange Indicator*—Prepare a solution containing 2 g of methyl orange per litre of 95 % ethyl alcohol.

19.5 *Procedure:*

19.5.1 Weigh a 5.0-g sample of the cement or a 10.0-g sample of concrete into a 250-mL beaker (Note 69). Disperse the sample with 75-mL of water. Without delay slowly add 25 mL of dilute (1+1) nitric acid, breaking up any lumps with a glass rod. If the smell of hydrogen sulfide is strongly evident at this point, add 3 mL of hydrogen peroxide (30 % solution) (Note 70). Add 3 drops of methyl orange indicator and stir. Cover the beaker with a watch glass and allow to stand for 1 to 2 min. If a yellow to yellow-orange color appears on top of the settled solids, the solution is not sufficiently acidic. Add additional dilute nitric acid (1+1) dropwise while stirring until a faint pink or red color persists. Then add 10 drops in excess. Heat the covered beaker rapidly to boiling. Do not allow to boil for more than a few seconds. Remove from the hot plate (Note 71).

NOTE 69—Use a 5-g sample for cement, and 10 g for concrete and other materials having an expected chloride content of less than about 0.15 % Cl. Use proportionally smaller samples for materials with higher chloride concentrations. Use cement and other powdered materials as is without grinding. Coarse samples require grinding to pass a 20-mesh sieve. If a sample is too fine, excessive silica gel may form during digestion with nitric acid, thereby slowing subsequent filtration.

NOTE 70—Slags and slag cements contain sulfide sulfur in concentrations that can interfere with the determination.

NOTE 71—It is important to keep the beaker covered during heating and digestion to prevent the loss of chloride by volatilization. Excessive amounts of acid should not be used since this results in early removal of the silver chloride coating from the silver billet electrode. A slurry that is only slightly acidic is sufficient.

19.5.2 Wash a 9-cm coarse-textured filter paper with four 25-mL increments of water using suction filtering provided by a 250-mL or 500-mL Büchner funnel and filtration flask. Discard the washings and rinse the flask once with a small portion of water. Reassemble the suction apparatus and filter the sample solution. Rinse the beaker and the filter paper twice with small portions of water. Transfer the filtrate from the flask to a 250-mL beaker and rinse the flask once with water. The original beaker may be used (Note 72). Cool the filtrate to room temperature. The volume should not exceed 175 mL.

NOTE 72—It is not necessary to clean all the slurry residue from the sides of the beaker nor is it necessary that the filter remove all of the fine material. The titration may take place in a solution containing a small amount of solid matter.

19.5.3 For instruments equipped with dial readout it is necessary to establish an approximate "equivalence point" by immersing the electrodes in a beaker of water and adjusting the instrument to read about 20 mV lower than midscale. Record the approximate millivoltmeter reading. Remove the beaker and wipe the electrodes with absorbent paper.

19.5.4 To the cooled sample (Note 73) beaker from 19.5.2, carefully pipet 2.00 mL of standard 0.05 N NaCl solution. Place the beaker on a magnetic stirrer and add a TFE-fluorocarbon-coated magnetic stirring bar. Immerse the electrodes into the solution taking care that the stirring bar does not strike the electrodes; begin stirring gently. Place the delivery tip of the 10-mL buret, filled to the mark with standard 0.05 N silver nitrate solution, in (preferably) or above the solution (Note 74).

NOTE 73—It is advisable to maintain constant temperature during measurement, for the solubility relationship of silver chloride varies markedly with temperature at low concentrations.

NOTE 74—If the tip of the buret is out of the solution, any adhering droplet should be rinsed onto the beaker with a few millilitres of water following each titration increment.

19.5.5 Gradually titrate, record the amount of standard 0.05 N silver nitrate solution required to bring the millivoltmeter reading to −60.0 mV of the equivalence point determined in the water.

19.5.6 Continue the titration with 0.20-mL increments. Record the buret reading and the corresponding millivoltmeter reading in columns 1 and 2 of a four-column recording form like that shown in Appendix X1. Allow sufficient time between each addition for the electrodes to reach equilibrium with the sample solution. Experience has shown that acceptable readings are obtained when the minimum scale reading does not change within a 5-s period (usually within 2 min).

19.5.7 As the equivalence point is approached, the equal additions of $AgNO_3$ solution will cause larger and larger changes in the millivoltmeter readings. Past the equivalence point the change per increment will again decrease. Continue to titrate until three readings past the approximate equivalence point have been recorded.

19.5.8 Calculate the difference in millivolt readings between successive additions of titrant and enter the values in Column 3 of the recording form. Calculate the difference between consecutive values in Column 3 and enter the results in Column 4. The equivalence point of the titration will be within the maximum Δ mV interval recorded in Column 3. The precise equivalence point can be interpolated from the data listed in Column 4 as shown in the Appendix X1.

19.5.9 *Blank*—Make a blank determination using 75 mL of water in place of the sample, following the same procedure starting with the fifth sentence of 19.5.1. Correct the results obtained in the analysis accordingly (Note 75) by subtracting the blank.

NOTE 75—For nonreferee analysis the blank may be omitted. In such case calculate the percent chloride in the sample using the following equation:

$$Cl, \% = 3.5453 \, (V\,N - 0.10)/W$$

where:

V = millilitres of 0.05 N $AgNO_3$ solution used for sample titration (equivalence point),
N = exact normality of 0.05 N $AgNO_3$ solution,
0.10 = milliequivalents of NaCl added (2.0 mL × 0.05 N), and
W = weight of sample, g.

19.6 *Calculations*—Calculate the percent chloride to the nearest 0.001 % as follows:

$$Cl, \% = 3.5453 \times V\,N/W$$

where:

V = millilitres of 0.05 N $AgNO_3$ solution used for titration of the sample (equivalence point),
N = exact normality of 0.05 N $AgNO_3$ solution, and
W = weight of sample, g.

20. Chloroform-Soluble Organic Substances (*Reference Method*)

20.1 *Summary of Method*—This method[10] was specially designed for the determination of

---

[10] The 1965 revision of these methods deleted the methoxyl method for determining Vinsol resin. Those interested in this method should refer to the *1966 Book of ASTM Standards*, Part 9.

Vinsol resin and tallow in portland cement, although mineral oil, common rosin, calcium stearate, and other fatty acid compounds, and probably some other substances, if present, will be included in the determination. Extreme care is necessary in the entire procedure. The method may be applied to types of cement other than portland cement, although if the cement contains a large amount of acid-insoluble matter, the emulsions may separate slowly, and less vigorous shaking, more chloroform, and more washing may be necessary.

20.2 *Reagents:*

20.2.1 *Chloroform*—If the blank determination as described in 20.3.5 exceeds 0.0015 g, the chloroform should be distilled before use. Chloroform recovered in the procedure may be slightly acid but can be reused for the portions to be shaken with the aqueous acid solution of the sample in the 1-L funnel. Chloroform used for washing the filter and transferring the extract should be fresh or distilled from fresh chloroform.

20.2.2 *Stannous Chloride* ($SnCl_2$).

20.3 *Procedure:*

20.3.1 Place 40 g of cement in a 1-L Squibb separatory funnel (Note 76) and mix it with 520 mL of water added in two approximately equal portions. Shake vigorously immediately after the addition of the first portion to effect complete dispersion. Then add the second portion and shake again. At once add rapidly 185 mL of HCl in which 10 g of $SnCl_2$ (Note 77) have been dissolved, rapidly insert the stopper in the funnel, invert, and shake with a swirling motion for a few seconds to loosen and disperse all the cement, taking care to avoid the development of great internal pressure due to unnecessarily violent shaking. Release internal pressure immediately by opening and closing the stopcock. Repeat the shaking and release the pressure until the decomposition of the cement is complete. If necessary, break up persistent lumps with a long glass rod. Cool to room temperature rapidly by allowing tap water to run on the flask.

NOTE 76—The use of grease to lubricate the stopcocks and glass stoppers of the separatory funnels should be avoided. Wetting the stopcocks with water before using will assist in their easy operation.

NOTE 77—The purpose of the $SnCl_2$ is to prevent the oxidation of sulfide sulfur to elemental sulfur, which is soluble in chloroform.

20.3.2 Add 75 mL of chloroform to the solution, stopper the funnel, shake it vigorously for 5 min, and allow the water and chloroform to stand 15 min to separate. Draw off the lower chloroform layer into a 125-mL Squibb separatory funnel, including the scum (Note 78) and a few millilitres of the aqueous layer, making certain that all the scum is transferred. Keep the amount of the aqueous layer transferred to an absolute minimum, since excessive water in the 125-mL funnel may result in incomplete extraction of the scum and may cause an emulsion which does not separate readily. Shake the funnel vigorously to ensure the complete extraction of the scum. Allow the chloroform to separate, and draw it into a 250-mL Squibb separatory funnel which contains 50 mL of water and a few drops of HCl, making sure to keep the scum behind in the 125-mL funnel. Shake the 250-mL funnel, and draw the chloroform into another 250-mL funnel that contains 50 mL of water and a few drops of HCl. Shake this funnel as in the case of the first 250-mL funnel. When the chloroform separates, draw it into a standard-taper flat-bottom boiling flask (Note 79), taking care not to allow any water to enter the flask.

NOTE 78—There is usually a dark colored scum at the liquid interface. It may contain chloroform-soluble organic substance after shaking in the funnel, where the proportion of water to chloroform is great. It may be concentrated and confined to a small volume by gently twirling the funnel after the scum has been drawn into the narrower part of the funnel.

NOTE 79—The liquid is later distilled. No cork or rubber stoppers should be used. A 250 or 300-mL soil analysis flask, fitted with a condenser tube by means of a ground joint, is satisfactory. The tube may be bent near the neck and the remaining part fitted with a water-cooling jacket. Chloroform thus recovered may be reused as described in 20.2.1.

20.3.3 Add 25 mL of chloroform to the solution in the original 1-L separatory funnel, and carry out the operations as described in 20.3.2, retaining the original wash water in the 250-mL funnels. Repeat, using another 25-mL portion of chloroform.

20.3.4 Distill the combined chloroform extracts in the boiling flask until their volume is reduced to 10 to 15 mL. Filter the remaining liquid into a weighed 100-mL glass beaker or platinum dish (Note 80) through a small medium-textured filter paper that has been washed with fresh chloroform. Rinse the flask and wash the paper with several small portions of fresh chloroform. Evaporate the extracts at a low temperature (not over 63°C) to dryness (Note 81) and heat it in an oven at 57 to 63°C for 3 min. Pass dry air into the vessel for 15 s, cool, and weigh. Repeat the heating and weighing until two successive weighings do not differ by more than 0.0010 g. The higher of the last two weights shall be taken as the true weight.

NOTE 80—A platinum dish is preferable, as it quickly attains the temperature of the balance. If a glass beaker is used, it should be allowed to stand in the case of the balance for at least 20 min before weighing.

NOTE 81—Care should be taken in drying the extract, as many of the chloroform-soluble organic substances are somewhat volatile when heated for a long time at even moderate temperatures. With protection from the accumulation of dust, the solution may be evaporated at room temperature overnight.

When a quick evaporation is desired, the solution may be evaporated on a hot plate at low heat under a stream of dry air through a glass tube (about 10 mm in inside diameter) until it is about 3 mm in depth. Then remove the vessel from the hot plate and continue a slow stream of dry air until the residue appears dry. Then continue with a more rapid stream of dry air for 5 min at room temperature before placing the vessel in the oven at 57 to 63°C. After each 3-min heating period in the oven, pass dry air into the vessel for about 15 s before weighing. The air may be dried by passing it through a cheap desiccant, such as calcium chloride or sulfuric acid, followed by a desiccant of high efficiency, such as magnesium perchlorate or anhydrous calcium sulfate, with care taken to avoid the carrying of dust from the desiccant by the air. Instead of using compressed air, which is often contaminated with oil, dirt, and moisture, one can place the chloroform solution under a bell glass and induce a stream of air through the desiccants by means of an aspirator or vacuum pump.

When Vinsol resin is known to be the only substance present, the residue is more stable and may be heated at 100 to 105°C, instead of 57 to 63°C, in order to expel all possible traces of chloroform.

20.3.5 *Blank*—Make a blank determination. Ignite a 40-g sample of the cement at 950 to 1000°C for 1 h (Note 82) and regrind. Treat this ignited sample by the same procedure and using the same reagents as in the analysis and correct the results accordingly.

NOTE 82—Care should be taken to completely burn off the organic substance. A 100-mL flat platinum dish, in which the sample is well spread out, and a muffle furnace are advised for this purpose. If such a furnace is not available, a large high-temperature burner of the Meker type may be used. Thorough stirring of the sample should be done frequently—every 5 min when a burner is used.

20.4 *Calculation*—Calculate the percentage of chloroform-soluble organic substances to the nearest 0.001 by multiplying the weight in grams of residue (Note 83) by 2.5 (100 divided by the weight of the sample used (40 g)).

NOTE 83—If the organic substance in the cement is tallow, the residue is the fatty acids resulting from the hydrolysis of the tallow in the hot acid solution, and its weight should be multiplied by 1.05 to give the weight of the original glycerides in the tallow. If the original substance is calcium stearate, the residue is stearic acid, and its weight multiplied by 1.07 gives the weight of calcium stearate.

ALTERNATE METHODS

21. Calcium Oxide (*Alternate Method*)

21.1 *Summary of Method:*

21.1.1 This method covers the gravimetric determination of CaO after removal of SiO_2 and the ammonium hydroxide groups and double precipitation of calcium as the oxalate. The precipitate is converted to CaO by ignition and is weighed.

21.1.2 Strontium, usually present in portland cement as a minor constituent, is precipitated with calcium as the oxalate and is subsequently calculated as CaO. If the SrO content is known and correction of CaO for SrO is desired as, for example, for research purposes or to compare results with SRM certificate values, the CaO obtained by this method may be corrected by subtracting percent SrO. In determining conformance of a cement to specifications the correction of CaO for SrO should not be made.

21.2 *Procedure* (Note 84):

21.2.1 Acidify the combined filtrates obtained in the determination of the ammonium hydroxide group (7.1 through 7.3) and, if necessary, evaporate to a volume of about 200 mL. Add 5 mL of HCl, a few drops of methyl red indicator solution, and 30 mL of warm ammonium oxalate solution (50 g/L) (Note 40). Heat the solution to 70 to 80°C and add NH_4OH (1+1) dropwise with stirring until the color changes from red to yellow (see Note 41). Allow the solution to stand without further heating for 1 h (not longer), with occasional stirring during the first 30 min. Filter using a retentive paper and wash moderately with cold ammonium oxalate solution (1 g/L). Reserve the filtrate and washings.

NOTE 84—When analyses are being made for determining conformity to specifications and there is a possibility that sufficient manganese will be present to cause the percentage of magnesium determined by alternate methods to exceed the specification limit, manganese may be removed as directed in 13.3.2 before CaO is determined by this alternate procedure.

21.2.2 Transfer the precipitate and filter paper to the beaker in which the precipitation was made. Dissolve the oxalate in 50 mL of hot HCl (1+4) and macerate the filter paper. Dilute to 200 mL with water, add a few drops of methyl red indicator and 20 mL of ammonium oxalate solution, heat the solution nearly to boiling, and precipitate calcium oxalate again by neutralizing the acid solution with NH_4OH as described in 13.3.1. Allow the solution to stand 1 to 2 h (standing for 2 h at this point does no harm), filter, and wash as before. Combine the filtrate with that already obtained and reserve for the determination of MgO (14.3.1).

21.2.3 Dry the precipitate in a weighed covered platinum crucible. Char the paper without inflaming, burn the carbon at as low a temperature as possible, and, finally, heat with the crucible tightly covered in an electric furnace or over a blast lamp at a temperature of 1100 to 1200°C. Cool in a desiccator and weigh as CaO. Repeat the ignition to constant weight.

21.2.4 *Blank*—Make a blank determination, following the same procedure and using the same amounts of reagents, and correct the results obtained in the analysis accordingly.

21.3 *Calculation:*

21.3.1 Calculate the percentage of CaO to the nearest 0.1 by multiplying the weight in grams of CaO by 200 (100 divided by the weight of sample used (0.5 g)).

21.3.2 Correct the percent CaO for SrO, if desired, by subtracting the percent SrO.

22. Magnesium Oxide (*Alternate Method*)

22.1 *Summary of Method*—The alternate method is a volumetric procedure suitable for use when the determinations of silicon dioxide (SiO_2), aluminum oxide (Al_2O_3), ferric oxide (Fe_2O_3), and calcium oxide (CaO) are omitted.

22.2 *Rapid Volumetric Method (Titration of Magnesium Oxyquinolate):*

22.3 *Reagents:*

22.3.1 *Ammonium Nitrate Solution* (20 g NH_4NO_3/L).

22.3.2 *Ammonium Oxalate Solution* (50 g/L).

22.3.3 *Hydroxyquinoline Solution*—Dissolve 25 g of 8-hydroxyquinoline in 60 mL of acetic acid. When the solution is complete, dilute to 2 L with cold water. One millilitre of this solution is equivalent to 0.0016 g of MgO.

22.3.4 *Potassium Bromate-Potassium Bromide, Standard Solution* (0.2 *N*)—Dissolve 20 g of potassium bromide (KBr) and 5.57 g of potassium bromate ($KBrO_3$) in 200 mL of water and dilute to 1 L. Obtain the ratio of the strength of this solution to that of the 0.1 *N* $Na_2S_2O_3$ solution (22.2.6) as follows: To 200 mL of water in a 500-mL Erlenmeyer flask add 25.0 mL of the 0.2 *N* $KBrO_3$–KBr solution, measured from a pipet or buret. Add 20 mL of HCl, stir, and add immediately 10 mL of potassium iodide (KI) (250 g/L). Mix well and titrate at once with the $Na_2S_2O_3$ solution until nearly colorless. Add 2 mL of starch solution and titrate to the disappearance of the blue color. Calculate the ratio in strength of the $KBrO_3$–KBr solution to the $Na_2S_2O_3$ solution by dividing the volume of $Na_2S_2O_3$ solution by the volume of $KBrO_3$–KBr solution used in the titration.

22.3.5 *Potassium Iodide Solution* (250 g KI/L).

22.3.6 *Sodium Thiosulfate, Standard Solution* (0.1 *N*)—Dissolve 25 g of sodium thiosulfate ($Na_2S_2O_3 \cdot 5H_2O$) in 200 mL of water, add 0.1 g of sodium carbonate (Na_2CO_3), and dilute to 1 L. Let stand at least 1 week. Standardize this solution directly against primary standard potassium dichromate ($K_2Cr_2O_7$). One millilitre of 0.10 *N* $Na_2S_2O_3$ solution is equivalent to 0.000504 g of MgO.

22.3.7 *Starch Solution*—To 500 mL of boiling water add a cold suspension of 5 g of soluble starch in 25 mL of water, cool to room temperature, add a cool solution of 5 g of sodium hydroxide (NaOH) in 50 mL of water, add 15 g of KI, and mix thoroughly.

22.4 *Procedure:*

22.4.1 Disperse 0.5 g (Note 85) of the sample of cement in a 400-mL beaker with 10 mL of water, using a swirling motion. While still swirling, add 10 mL of HCl all at once. Dilute immediately to 100 mL. Heat gently and grind any coarse particles with the flattened end of a glass rod until decomposition is complete, add 2 or 3 drops of HNO_3 and heat to boiling (Note 86).

NOTE 85—If SiO_2, ammonium hydroxide group, and CaO are separated and determined in accordance with the appropriate sections for either the reference or alternate methods, the remaining filtrate may be used for the determination of MgO as described in 22.4.1, starting with the third from the last sentence of 22.4.2, "Add 5 mL of HCl...".

NOTE 86—In the case of cements containing blast-furnace slag or a significant quantity of sulfide sulfur, add 12 drops of HNO_3 and boil for 20 min to oxidize iron and remove sulfide.

22.4.2 Add 3 drops of methyl red indicator to the solution and then add NH_4OH until the solution is distinctly yellow. Heat this solution to boiling and boil for 50 to 60 s. In the event difficulty from bumping is experienced while boiling the ammoniacal solution, a digestion period of 10 min on a steam bath, or a hot plate having the approximate temperature of a steam bath, may be substituted for the 50 to 60-s boiling period. Remove from the burner, steam bath, or hot plate and allow to stand until the precipitate has settled. Using medium-textured paper, filter the solution without delay, wash the precipitate twice with hot NH_4NO_3 (20 g/L), and reserve the filtrate. Transfer the precipitate with the filter paper to the beaker and dissolve in 10 mL of HCl (1+1). Macerate the filter paper. Dilute to about 100 mL and heat to boiling. Reprecipitate, filter, and wash the hydroxides as above. Combine this filtrate and washings with those from the first precipitation taking care that the volume does not exceed 300 mL (Note 87). Add 5 mL of HCl, a few drops of methyl red indicator solution and 30 mL of warm ammonium oxalate solution (50 g/L). Heat the solution to 70 to 80°C and add NH_4OH (1+1) dropwise, while stirring, until the color changes from red to yellow (see Note 41). Allow the solution to stand without further heating for 15 min on a steam bath.

NOTE 87—In the case of cements containing blast-furnace slag, or which are believed to contain a significant quantity of manganese, acidify with HCl, evaporate to about 100 mL, and remove the manganese, using the procedure described in 13.3.1.

22.4.3 Add 10 to 25 mL of the 8-hydroxyquinoline reagent (Note 88) and then 4 mL of $NH_4OH/100$ mL of solution. Stir the solution on a mechanical stirring machine for 15 min and set aside until the precipitate has settled (Note 89). Filter the solution using medium-textured paper and wash the precipitate with hot NH_4OH (1+40). Dissolve the precipitate in 50 to 75 mL of hot HCl (1+9) in a 500-mL Erlenmeyer flask. Dilute the resulting solution to 200 mL and add 15 mL of HCl. Cool the solution to 25°C and add 10 to 35 mL of the 0.2 N $KBrO_3$–KBr solution (Note 90) from a pipet or buret. Stir the solution and allow to stand for about 30 s to ensure complete bromination. Add 10 mL of KI (250 g/L). Stir the resulting solution well and then titrate with the 0.1 N $Na_2S_2O_3$ solution until the color of the iodine becomes faintly yellow. At this point add 2 mL of the starch solution and titrate the solution to the disappearance of the blue color.

NOTE 88—An excess of the 8-hydroxyquinoline reagent is needed to avoid a low result for MgO, but too great an excess will yield high results. The following guide should be used to determine the amount of reagent added:

Approximate Content of MgO, %	Approximate Amount of Reagent Required, mL
0 to 1.5	10
1.5 to 3.0	15
3.0 to 4.5	20
4.5 to 6.0	25

NOTE 89—The precipitate should be filtered within an hour. Prolonged standing may cause high results.

NOTE 90—The amount of the standard $KBrO_3$–KBr solution used should be as follows:

Approximate Content of MgO, %	Amounts of Standard $KBrO_3$ – KBr Solution, mL
0 to 1	10
1 to 2	15
2 to 3	20
3 to 4	25
4 to 5	30
5 to 6	35

22.4.4 *Blank*—Make a blank determination, following the same procedure and using the same amounts of reagents, and correct the results obtained in the analysis accordingly.

22.5 *Calculation*—Calculate the percentage of MgO to the nearest 0.1 as follows: (Note 91)

$$MgO, \% = E(V_1R - V_2) \times 200$$

where:
E = MgO equivalent of the $Na_2S_2O_3$ solution, g/mL,
V_1 = millilitres of $KBrO_3$–KBr solution used,
R = ratio in strength of the $KBrO_3$–KBr solution to the $Na_2S_2O_3$ solution,
V_2 = millilitres of $Na_2S_2O_3$ solution used, and
200 = 100 divided by the weight of sample used (0.5 g).

NOTE 91—V_1R represents the volume of $Na_2S_2O_3$ solution equivalent to the volume of $KBrO_3$–KBr solution used. V_2 represents the amount of $Na_2S_2O_3$ required by the excess $KBrO_3$–KBr which is not reduced by magnesium oxyquinolate.

23. Loss on Ignition

23.1 *Portland Blast Furnace Slag Cement and Slag Cement (Alternate Method)*:

23.1.1 *Summary of Method*—This method

covers a correction for the gain in weight due to oxidation of sulfides usually present in such cement by determining the decrease in the sulfide sulfur content during ignition. It gives essentially the same result as the reference method (16.2.1 through 16.2.3) which provides for applying a correction based on the increase in SO_3 content.

23.1.2 *Procedure:*

23.1.2.1 Weigh 1 g of cement in a tared platinum crucible, cover, and ignite in a muffle furnace at a temperature of 950 ± 50°C for 15 min. Cool to room temperature in a desiccator and weigh. After weighing carefully transfer the ignited material to a 500-mL boiling flask. Break up any lumps in the ignited cement with the flattened end of a glass rod.

23.1.2.2 Determine the sulfide sulfur content of the ignited sample using the procedure described in 15.2.1 through 15.2.5. Using the same procedure, also determine the sulfide sulfur content of a portion of the cement that has not been ignited.

23.1.3 *Calculation:*

23.1.3.1 Calculate the percentage loss of weight occurring during ignition (23.1.2.1) and add twice the difference between the percentages of sulfide sulfur in the original sample and ignited sample as determined in 23.1.2.2. Report this value as the loss on ignition.

NOTE 92—If a gain of weight is obtained during the ignition, subtract the percentage of gain from the correction for sulfide oxidation.

24. Titanium Dioxide (*Alternate Method*)

24.1 *Summary of Method*—In this method, titanium dioxide (TiO_2) is determined colorimetrically by comparing the color intensity of the peroxidized solution of the titanium in the sample with the color intensity of a peroxidized standard solution of titanic sulfate.

24.2 *Interferences*—Interfering elements in the peroxide method for TiO_2 are vanadium, molybdenum, and chromium. In very small quantities the interference of the last two is negligible. However, vanadium in very small quantities causes interference and, as some cements contain this element, the Na_2CO_3 fusion (24.5.4) and extraction with water are necessary.

24.3 *Apparatus:*

24.3.1 *Colorimeter*—The apparatus shall consist of a colorimeter of the Kennicott or Duboscq type, or other colorimeter or spectrophotometer designed to measure light transmittancy, and suitable for measurements at wavelengths between 400 and 450 nm.

24.4 *Reagents:*

24.4.1 *Ammonium Chloride* (NH_4Cl).

24.4.2 *Ammonium Nitrate* (20 g NH_4NO_3/L).

24.4.3 *Ferrous Sulfate Solution* (1 mL = 0.005 g Fe_2O_3)—Dissolve 17.4 g of ferrous sulfate ($FeSO_4 \cdot 7H_2O$) in water containing 50 mL of H_2SO_4 and dilute to 1 L. One millilitre is equivalent to 1 % of Fe_2O_3 in 0.5 g of sample.

24.4.4 *Hydrogen Peroxide* (30 %)—Concentrated hydrogen peroxide (H_2O_2).

24.4.5 *Sodium Carbonate* (20 g Na_2CO_3/L).

24.4.6 *Sodium or Potassium Pyrosulfate* ($Na_2S_2O_7$ or $K_2S_2O_7$).

24.4.7 *Titanic Sulfate, Standard Solution* (1 mL = 0.0002 g TiO_2)—Use standard TiO_2 furnished by the National Bureau of Standards (Standard Sample 154 or its replacements). Dry for 2 h at 105 to 110°C. Transfer a weighed amount, from 0.20 to 0.21 g of the TiO_2 to a 125-mL Phillips beaker. Add 5 g of ammonium sulfate (($NH_4)_2SO_4$) and 10 mL of H_2SO_4 to the beaker and insert a short-stem glass funnel in the mouth of the beaker. Heat the mixture cautiously to incipient boiling while rotating the flask over a free flame. Continue the heating until complete solution has been effected and no unattacked material remains on the wall of the flask (Note 93). Cool and rapidly pour the solution into 200 mL of cold water while stirring vigorously. Rinse the flask and funnel with H_2SO_4 (1+19), stir, and let the solution and washings stand for at least 24 h. Filter into a 1-L volumetric flask, wash the filter thoroughly with H_2SO_4 (1+19), dilute to the mark with H_2SO_4 (1+19), and mix.

NOTE 93—There may be a small residue, but it should not contain more than a trace of TiO_2 if the operations have been properly performed.

24.4.8 Calculate the TiO_2 equivalent of the titanic sulfate solution, g/mL, as follows:

$$E = AB/1000$$

where:

E = TiO_2 equivalent of the $Ti(SO_4)_2$ solution, g/mL,

A = grams of standard TiO_2 used (corrected for loss on drying),

B = percentage of TiO_2 in the standard TiO_2

as certified by the National Bureau of Standards, divided by 100, and
1000 = number of millilitres in the volumetric flask.

24.5 *Procedure:*

24.5.1 Mix thoroughly 0.5 g of the sample of cement and about 0.5 g of NH_4Cl in a 50-mL beaker, cover the beaker with a watch glass, and add cautiously 5 mL of HCl, allowing the acid to run down the lip of the covered beaker. After the chemical action has subsided, lift the cover, stir the mixture with a glass rod, replace the cover, and set the beaker on a steam bath for 30 min (Note 94). During this time of digestion, stir the contents occasionally and break up any remaining lumps to facilitate the complete decomposition of the cement. Fit a medium-textured filter paper to a funnel and transfer the precipitate to the filter. Scrub the beaker with a rubber policeman and rinse the beaker and policeman. Wash the filter two or three times with hot HCl (1+99) and then with ten or twelve small portions of hot water, allowing each portion to drain through completely.

NOTE 94—A hot plate may be used instead of a steam bath if the heat is so regulated as to approximate that of a steam bath.

24.5.2 Transfer the filter and residue to a platinum crucible (Note 95), dry, and ignite slowly until the carbon of the paper is completely consumed without inflaming. Treat the SiO_2 thus obtained with 0.5 to 1 mL of water, about 10 mL of HF, and 1 drop of H_2SO_4, and evaporate cautiously to dryness (Note 96).

NOTE 95—When it is desired to shorten the procedure for purposes other than referee analysis, usually with little sacrifice of accuracy, the procedure given in 24.5.2 may be omitted.

NOTE 96—When a determination of SiO_2 is desired in addition to one of TiO_2, the SiO_2 may be obtained and treated with HF as directed in 6.2.3.1 through 6.2.4.

24.5.3 Heat the filtrate to boiling and add NH_4OH until the solution becomes distinctly alkaline, as indicated by an ammoniacal odor. Add a small amount of filter paper pulp to the solution and boil for 50 to 60 s. Allow the precipitate to settle, filter through a medium-textured paper, and wash twice with hot NH_4NO_3 solution (20 g/L). Place the precipitate in the platinum crucible in which the SiO_2 has been treated with HF and ignite slowly until the carbon of the paper is consumed.

NOTE 97—When a determination of ammonium hydroxide group is desired in addition to one of TiO_2, the precipitation and ignition may be made as described in 7.2.1 through 7.2.4. However, the crucible must contain the residue from the treatment of the SiO_2 with HF unless circumstances permit its omission as indicated in Note 95.

24.5.4 Add 5 g of Na_2CO_3 to the crucible and fuse for 10 to 15 min (see 24.2.1). Cool, separate the melt from the crucible, and transfer to a small beaker. Wash the crucible with hot water, using a policeman. Digest the melt and washings until the melt is completely disintegrated, then filter through a 9-cm medium-textured filter paper and wash a few times with Na_2CO_3 (20 g/L). Discard the filtrate. Place the precipitate in the platinum crucible and ignite slowly until the carbon of the paper is consumed.

24.5.5 Add 3 g of $Na_2S_2O_7$ or $K_2S_2O_7$ to the crucible and heat below red heat until the residue is dissolved in the melt (Note 98). Cool and dissolve the fused mass in water containing 2.5 mL of H_2SO_4. If necessary, reduce the volume of the solution (Note 99), filter into a 100-mL volumetric flask through a 7-cm medium-textured filter paper, and wash with hot water. Add 5 mL of H_3PO_4, and cool the solution to room temperature. Add H_2O_2 (1.0 mL of 30 % strength or its equivalent) (Note 100), dilute to the mark with water, and mix thoroughly.

NOTE 98—Start the heating with caution because pyrosulfates (also known as fused bisulfates) as received often foam and spatter in the beginning due to an excess of H_2SO_4. Avoid an unnecessarily high temperature or unnecessarily prolonged heating, as fused pyrosulfates may attack platinum. A supply of nonspattering pyrosulfates may be prepared by heating some pyrosulfate in a platinum vessel to eliminate the excess H_2SO_4 and crushing the cool fused mass.

NOTE 99—If the solution is evaporated to too small a volume and allowed to cool, there may be a precipitate of sulfates difficult to redissolve. In case of over-evaporation, do not permit the contents to cool, but add hot water and digest on a steam bath or hot plate until the precipitate is redissolved, with the possible exception of a small amount of SiO_2.

NOTE 100—Hydrogen peroxide deteriorates on standing. Its strength may be determined by adding a measured volume of the solution to 200 mL of cold water and 10 mL of H_2SO_4 (1+1) and titrating with a standard solution of potassium permanganate ($KMnO_4$) prepared in accordance with 13.2.2. If the standard solution contains 0.0056357 g of $KMnO_4$/mL, 49.5 mL of it will be required by 0.50 mL of H_2O_2 (30 %).

24.5.6 Prepare from the standard $Ti(SO_4)_2$ solution a suitable reference standard solution or a series of reference standard solutions in 100-mL volumetric flasks, depending upon the type of colorimeter to be used. To each solution add 3 g of $Na_2S_2O_7$ or $K_2S_2O_7$ dissolved in water, an amount of $FeSO_4$ solution equivalent to the Fe_2O_3 content in 0.5 g of the cement under test, 2.5 mL of H_2SO_4, and 5 mL of H_3PO_4 (Note 101). When the solution is at room temperature, add H_2O_2 (1.0 mL of 30 % strength or its equivalent), dilute to the mark with water, and mix thoroughly (Note 102).

NOTE 101—The color imparted to the solution by $Fe_2(SO_4)_2$ is partly offset by the bleaching effect of H_2SO_4, H_3PO_4, and alkali salts on ferric and peritanic ions. The directions should be followed closely for the highest degree of precision. However, when it is desired to shorten this procedure for purposes other than referee analysis, the addition of pyrosulfate, $FeSO_4$ solution and H_3PO_4 to the color comparison solutions may be omitted provided the Fe_2O_3 of the sample cement is less than 5 %. This usually leads to little sacrifice to accuracy.

NOTE 102—The color develops rapidly and is stable for a sufficient period of time, but if the peroxidized solution is allowed to stand a long time, bubbles of oxygen may appear and interfere with color comparison. When the contents of a tube are first mixed, there may be fine bubbles which should be allowed to clear up before the comparison is made. Comparison between the standard and unknown solution should be made not less than 30 min after addition of H_2O_2.

24.5.7 Compare the color, light transmittancy, or absorbance of the unknown solution with the reference standard solution. The technique of comparing colored solutions or measuring transmittancy or absorbance depends on the type of apparatus (see 24.5.8 to 24.5.10) and should be in accordance with standard practice appropriate to the particular type used or with instructions supplied by the manufacturer of the equipment. If the peroxidized solution of cement is compared with a single standard peroxidized solution, bear in mind that a single peroxidized solution cannot be used for the whole range in TiO_2 content that may be encountered. The difference in volume or depth for the two liquids should not exceed 50 % of the smaller value. All solutions should contain the prescribed concentrations of H_2SO_4, H_3PO_4, $Fe_2(SO_4)_3$, and persulfate except under the circumstances indicated in Note 101.

24.5.8 *Colorimeter of the Kennicott Type*—By means of a plunger in a reservoir of standard peroxidized solution, adjust the amount of solution through which light passes until it gives the same color intensity as the peroxidized solution of the sample.

24.5.9 *Colorimeter of the Duboscq Type*—Lower or raise the plungers in the cups until the two solutions give the same color intensity when viewed vertically. The color matching may be done either visually or photoelectrically.

24.5.10 *Colorimeter Designed to Measure Light Transmittancy*—The measurement should be made between 400 to 450 nm and may be made either visually or photoelectrically. In most colorimeters of this type, the instrument is calibrated with standard solutions and a calibration curve showing the relation of light transmittancy or absorbance to TiO_2 content is prepared in advance of the analysis of the sample for TiO_2.

24.5.11 *Blank*—Make a blank determination, following the same procedure and using the same amounts of reagent, and correct the results obtained in the analysis accordingly.

24.6 *Calculation*—Calculate the percentage of TiO_2 to the nearest 0.01. When a colorimeter designed to measure light transmittancy is used, read the percentage of TiO_2 from a calibration curve showing the relation of light intensity to TiO_2 content. When the peroxidized solution of the sample is compared with a single reference standard solution, calculate the percentage of TiO_2 as follows (Note 103):

24.6.1 *For Colorimeters of the Kennicott Type:*

$TiO_2, \% = (100 \ VE/S) \times (D/C)$

24.6.2 *For Colorimeters of the Duboscq Type:*

$TiO_2, \% = (100 \ VE/S) \times (F/G)$

where:
V = millilitres of standard $Ti(SO_4)_2$ solution in the peroxidized standard solution,
E = TiO_2 equivalent of the standard $Ti(SO_4)_2$ solution, g/mL,
S = grams of sample used,
C = total volume of the peroxidized reference standard solution, mL,
D = volume of peroxidized reference standard solution that matches the peroxidized solution of the sample, mL,
F = depth of peroxidized reference standard solution through which light passes, and
G = depth of peroxidized solution of the sample through which light passes.

NOTE 103—The difference between D and C or between F and G should not exceed 50 % of the smaller value.

25. Phosphorus Pentoxide (*Alternate Method*)

25.1 *Summary of Method*—In this method, phosphorus is determined volumetrically by precipitation of the phosphorus as ammonium phosphomolybdate and titration with NaOH and H_2SO_4.

25.2 *Reagents:*

25.2.1 *Ammonium Molybdate Solution*—Prepare the solution in accordance with 9.3.1.

25.2.2 *Ammonium Nitrate* (NH_4NO_3).

25.2.3 *Potassium Nitrate Solution* (10 g/L)—Dissolve 10 g of potassium nitrate (KNO_3) in water freshly boiled to expel CO_2 and cooled, and dilute to 1 L.

25.2.4 *Sodium Hydroxide, Standard Solution* (0.3 N)—Dissolve 12 g of sodium hydroxide (NaOH) in 1 L of water that has been freshly boiled to expel CO_2, and cooled. Add 10 mL of a freshly filtered, saturated solution of barium hydroxide ($Ba(OH)_2$). Shake the solution frequently for several hours, and filter it. Protect it from contamination by CO_2 in the air. Standardize the solution against standard acid potassium phthalate (Standard Sample No. 84) or benzoic acid (Standard Sample No. 39) furnished by the National Bureau of Standards, according to the directions furnished with the standard. Calculate the phosphorus pentoxide (P_2O_5) equivalent (Note 104) of the solution, g/mL, as follows:

$$E = N \times 0.003086$$

where:

E = P_2O_5 equivalent of the NaOH solution, g/mL,

N = normality of the NaOH solution, and

0.003086 = P_2O_5 equivalent of 1 N NaOH solution, g/mL.

NOTE 104—The value of the solution is based on the assumption that the phosphorus in cement is precipitated as ammonium phosphomolybdate ($2(NH_4)_3PO_4 \cdot 12MoO_3$) and that the precipitate reacts with the NaOH solution thus:

$2(NH_4)_3PO_4 \cdot 12MoO_3 + 46NaOH$
$= 2(NH_4)_2HPO_4 + (NH_4)_2MoO_4$
$+ 23Na_2MoO_4 + 22H_2O$

The number of 0.003086 is obtained by dividing the molecular weight of P_2O_5 (141.96) by 46 (for 46 NaOH in the equation) and by 1000 (number of millilitres in 1 L).

As the actual composition of the precipitate is influenced by the conditions under which the precipitation is made, it is essential that all the details of the procedure are followed closely as prescribed.

25.2.5 *Sodium Nitrite* (50 g $NaNO_2$/L).

25.2.6 *Sulfuric Acid, Standard Solution* (0.15 N)—Dilute 4.0 mL of H_2SO_4 to 1 L with water that has been freshly boiled and cooled. Standardize against the standard NaOH solution. Determine the ratio in strength of the standard H_2SO_4 solution to the standard NaOH solution by dividing the volume of NaOH solution by the volume of H_2SO_4 solution used in the titration.

25.3 *Procedure:*

25.3.1 Weigh 1 to 3 g of the sample (Note 105) and 10 g of NH_4NO_3 into a 150-mL beaker. Mix the contents, add 10 mL of HNO_3, and stir quickly, using the flattened end of a glass rod to crush lumps of cement, until the cement is completely decomposed and the thick gel of silica (SiO_2) is broken up. Cover the beaker with a watch glass, place it on a water bath or a hot plate at approximately 100°C for 15 to 20 min, and stir the contents occasionally during the heating. Add 20 mL of hot water to the beaker and stir the contents. If the cement contains an appreciable amount of manganese, as shown by the presence of a red or brown residue, add a few millilitres of $NaNO_2$ (50 g/L) to dissolve this residue. Boil the contents of the beaker until all nitrous fumes are completely expelled. This procedure should not take more than 5 min, and water should be added to replace any lost by evaporation. Filter, using medium-textured paper, into a 400-mL beaker under suction and with a platinum cone to support the filter paper. Wash the residue of SiO_2 with hot water until the volume of filtrate and washings is about 150 mL.

NOTE 105—The amounts of sample and reagents used depend on the content of phosphorus in the cement. The minimum requirements are sufficient if the cement contains 0.5 % P_2O_5 or more. The maximum amounts are required if the content of P_2O_5 is 0.1 % or less.

25.3.2 Heat the solution to 69 to 71°C, remove it from the heat source, and immediately add 50 to 100 mL of the ammonium molybdate solution. Stir the solution vigorously for 5 min, wash down the sides of the beaker with cool KNO_3 solution (10 g/L), cover the beaker with a watch glass, and allow to stand 2 h. Using suction, filter the precipitate (Note 106), decanting the solution with as little disturbance to the precipitate as possible. Stir the precipitate in the beaker with a stream of the cool KNO_3 solution, decant the liquid, then wash the precipitate onto the filter. Scrub the stirring rod and beaker with a policeman and wash the contents onto the filter. Wash and precipitate until it is acid-free (Note 107), allowing each portion of wash solution to be sucked completely through before adding the next.

NOTE 106—The filter may be a small medium-textured filter paper supported by a platinum cone, or a small Hirsch funnel may be used with filter paper cut to fit and a thin mat of paper pulp or acid-washed asbestos pulp. The filtration should be carried out with care to avoid any loss of the precipitate. The filter should fit well, and the suction should be started before filtration and maintained until the end of the washing.

NOTE 107—About ten washings are usually required. Test the tenth washing with one drop of neutral phenolphthalein indicator and half a drop of the standard NaOH solution. If a definite pink color lasts at least 5 min, the precipitate is considered to be acid-free; otherwise, continue the washing.

25.3.3 Transfer the filter and precipitate to the beaker in which the precipitation took place, using small damp pieces of paper to wipe out the funnel and to pick up portions of the precipitate that may remain on it. Add 20 mL of cool CO_2-free water to the beaker, and break up the filter by stirring rapidly with the policeman that was used to scrub the beaker. Add an excess of the 0.3 N NaOH solution, stir the contents until all trace of yellow has disappeared, wash down the policeman and sides of the beaker with 50 mL of cool, CO_2-free water, and add 2 drops of neutral phenolphthalein indicator solution. Treat the solution with a measured quantity of the 0.15 N H_2SO_4 solution, sufficient to destroy completely the pink color. Complete the titration with the NaOH solution until there is a definite faint pink color that lasts at least 5 min.

25.3.4 *Blank*—Make a blank determination, following the same procedure and using the same amounts of reagents, and correct the results obtained in the analysis accordingly.

25.4 *Calculation*—Calculate the percentage of P_2O_5 to the nearest 0.01 as follows:

$$P_2O_5, \% = [E(V_1 - V_2R)/S] \times 100$$

where:
E = P_2O_5 equivalent of the NaOH solution, g/mL,
V_1 = millilitres of NaOH solution used,
V_2 = millilitres of H_2SO_4 solution used,
R = ratio in strength of the H_2SO_4 solution to the NaOH solution, and
S = grams of sample used.

26. Manganic Oxide (*Alternate Method*)

26.1 *Summary of Method*—In this method manganic oxide is determined volumetrically by titration with potassium permanganate solution.

26.2 *Reagents:*

26.2.1 *Potassium Permanganate, Standard Solution* (0.18 N)—Prepare a solution of potassium permanganate ($KMnO_4$) and standardize as described in 13.2.2, except that the manganic oxide (Mn_2O_3) equivalent of the solution is calculated instead of the calcium oxide (CaO) equivalent. Calculate the Mn_2O_3 equivalent of the solution as follows:

$$E = (B \times 0.3534)/A$$

where:
E = Mn_2O_3 equivalent of the $KMnO_4$ solution, g/mL,
B = grams of $Na_2C_2O_4$ used,
A = millilitres of $KMnO_4$ solution required by the $Na_2C_2O_4$, and
0.3534 = mole ratio of 3 Mn_2O_3 to 10 $Na_2C_2O_4$.

26.2.2 *Zinc Oxide* (ZnO), powder.

26.3 *Procedure:*

26.3.1 Place 2 g of the sample in a 250-mL beaker and add about 50 mL of water to the cement. Stir the mixture until it is in suspension and then add about 15 mL of HCl. Heat the mixture gently until the solution is as complete as possible. Add 5 mL of HNO_3 and 50 mL of water to the solution and boil it until most of the chlorine has been expelled. If necessary, add hot water to maintain the solution at a volume of about 100 mL. Stop the boiling and add ZnO powder to the solution until the acid is neutralized. Add an excess of 3 to 5 g of ZnO powder to the solution and boil it for a few minutes.

26.3.2 Without filtering, and while keeping the solution hot (90 to 100°C) by intermittent or continuous heating, titrate the solution with the 0.18 N $KMnO_4$ solution until a drop of it gives a permanent pink color (Note 108). When the end point is approached, add the standard solution dropwise. After each drop, stir the solution, allow the precipitate to settle a little, and observe the color of the stratum of the solution by looking through the side of the beaker.

NOTE 108—In the case of a cement in which the approximate content of Mn_2O_3 is unknown, a preliminary determination may be made with rapid titration, 0.5 to 1 mL of the standard solution being added at a time, and without an attempt to keep the solution close to the boiling point.

26.3.3 *Blank*—Make a blank determination, following the same procedure and using the same amounts of reagents, and correct the results obtained in the analysis accordingly.

26.4 *Calculation*—Calculate the percentage of Mn_2O_3 to the nearest 0.01 as follows:

$$Mn_2O_3, \% = EV \times 50$$

where:
- E = Mn_2O_3 equivalent of the $KMnO_4$ solution, g/mL,
- V = millilitres of $KMnO_4$ solution used, and
- 50 = 100 divided by the weight of sample used (2 g).

27. Free Calcium Oxide *(Alternate Methods)*

27.1 *Summary of Methods*—These are rapid methods for the determination of free calcium oxide in fresh clinker. When applied to cement or aged clinker, the possibility of the presence of calcium hydroxide should be kept in mind since these methods do not distinguish before free CaO and free $Ca(OH)_2$. Two methods are provided. Alternate Method A is a modified Franke procedure in which uncombined lime is titrated with dilute perchloric acid after solution in an ethylacetoacetate - isobutylalcohol solvent. Alternate Method B is an ammonium acetate titration of the alcohol - glycerin solution of uncombined lime with $Sr(NO_3)_2$ as an accelerator.

27.2 *Modified Franke Method (Alternate Method A):*

27.2.1 *Apparatus:*

27.2.1.1 *Refluxing Assembly,* consisting of a flask with flat-bottom short neck Erlenmeyer flask with 250-mL capacity. The water-cooled refluxing condenser should have a minimum length of 300 mm. The flask and reflux condenser shall be connected with standard tapered ground glass joints. The reflux condenser shall be fitted with an absorption tube containing a desiccant, such as indicating silica gel, and a material for the removal of CO_2, such as Ascarite. The absorption tube shall be inserted with a rubber stopper in the upper end of the reflux column.

27.2.1.2 *Buret,* having a 10-mL capacity and graduated in units not more than 0.05 mL.

27.2.1.3 *Vacuum Filtration Assembly,* consisting of a Gooch crucible size No. 3, 25-mL capacity in which is placed a suitable filter paper, 21-mm size, a Walter crucible holder, a 500-mL vacuum flask, and vacuum source. The crucible is half filled with compressed filter pulp.

27.2.1.4 *Glass Boiling Beads:*

27.2.2 *Solutions Required:*

27.2.2.1 *Ethyl Acetoacetate - Isobutyl Alcohol Solvent*—3 parts of volume of ethyl acetoacetate and 20 parts by volume of isobutyl alcohol.

27.2.2.2 *Thymol Blue Indicator*—Dissolve 0.1 g of thymol blue indicator powder in 100 mL of isobutyl alcohol.

27.2.2.3 *Perchloric Acid, Standard Solution* (0.2 *N*)—Dilute 22 mL of 70 to 72 % perchloric acid to 1 L with isobutyl alcohol. Standardize this solution as follows: Ignite 0.1000 g of primary standard calcium carbonate in a platinum crucible at 900 to 1000°C. Cool the crucible and contents in a desiccator and weigh to the nearest 0.0001 g to constant weight. Perform the weighings quickly to prevent absorption of water and CO_2. Immediately transfer the CaO without grinding to a clean, dry Erlenmeyer flask and reweigh the empty crucible to the nearest 0.0001 g to determine the amount of CaO added. Then follow procedure beginning with "Add 70 mL of the ethyl acetoacetate - isobutyl alcohol ..." in 27.2.3.1. Calculate the CaO equivalents of the standard perchloric acid solution in grams per millilitre by dividing the weight of CaO used by the volume of perchloric acid required for the titration.

27.2.3 *Procedure:*

27.2.3.1 Weigh 1.0000 g of ground sample (Note 109) and transfer it into a clean, dry 250-mL Erlenmeyer flask. Add four to five glass boiling beads. Add 70 mL of prepared ethyl acetoacetate - isobutyl alcohol solvent. Agitate the flask to disperse the sample.

NOTE 109—Thorough grinding of the sample is essential for proper exposure of the free lime grains that often are occluded in crystals of tricalcium silicate in the cement. However, exposure of the sample to the air must be kept at a minimum to prevent carbonation of the free lime. In particular, direct breathing into the sample must be avoided. The sample should be sufficiently fine to easily pass a No 200 (75-μm) sieve but actual sieving is not recommended. If the sample is not to be immediately tested, it must be kept in an airtight container to avoid unnecessary exposure to the atmosphere.

27.2.3.2 Attach the flask to a reflux condenser and bring the material to a boil. Reflux for 15 min.

27.2.3.3 Remove flask from condenser, stopper, and cool rapidly to room temperature.

27.2.3.4 Filter the sample and solution using the vacuum assembly. Wash the flask and residue with small increments (10 to 15 mL) of isobutyl alcohol until a total of 50 mL has been used for wash solution.

27.2.3.5 Add 12 drops of the thymol blue indicator to the filtrate and immediately titrate with 0.2 N perchloride acid to the first distinct color change.

27.2.4 *Calculations*—Calculate the percent free calcium oxide to the nearest 0.1 % as follows:

$$\text{free CaO, \%} = \frac{EV \times 100}{W}$$

where:

E = CaO equivalent of the perchloric acid, g/mL,
V = millilitres of perchloric acid solution required by sample, and
W = weight of the sample, g.

27.3 *Rapid $Sr(NO_3)_2$ Method (Alternate Method B):*

27.3.1 *Reagents:*

27.3.1.1 *Ammonium Acetate, Standard Solution* (1 mL = 0.005 g CaO)—Prepare a standard solution of ammonium acetate ($NH_4C_2H_3O_2$) by dissolving 16 g of desiccated ammonium acetate in 1 L of ethanol in a dry, clean, stoppered bottle. Standardize this solution by the same procedure as described in 27.3.2.1, except use the following in place of the sample: ignite to constant weight approximately 0.1 g of calcium carbonate ($CaCO_3$) in a platinum crucible at 900 to 1000°C, cool the contents in a desiccator, and weigh to the nearest 0.0001 g. Perform the weighings quickly to prevent absorption of water and CO_2. Immediately transfer the CaO without grinding to a 250-mL boiling flask (containing glycerin – ethanol solvent and $Sr(NO_3)_2$), and reweigh the empty crucible to determine the weight of CaO to the nearest 0.0001 g. Continue as described in 27.3.2.1 and 27.3.2.2. Calculate the CaO equivalent of the ammonium acetate in grams per millilitre by dividing the weight of CaO used by the volume of solution required.

27.3.1.2 *Phenolphthalein Indicator*—Dissolve 1.0 g of phenolphthalein in 100 mL of ethanol (Formula 2B) (Note 110).

27.3.1.3 *Glycerin - Ethanol Solvent (1+2)*—Mix 1 volume of glycerin with 2 volumes of ethanol (Formula 2B). To each litre of this solution, add 2.0 mL of phenolphthalein indicator solution.

NOTE 110—Ethanol denatured in accordance with Formula 2B (99.5 % ethanol and 0.5 % benzol) is preferred but may be replaced by isopropyl alcohol, A.R.

27.3.1.4 *Strontium Nitrate* ($Sr(NO_3)_2$), reagent grade.

27.3.2 *Procedure:*

27.3.2.1 Transfer 60 mL of the glycerin – ethanol solvent into a clean, dry, 250-mL standard-taper flat-bottom boiling flask. Add 2 g of anhydrous strontium nitrate ($Sr(NO_3)_2$), and adjust the solvent to slightly alkaline with a dropwise addition of a freshly prepared dilute solution of NaOH in ethanol until a faint pink color is formed. Weigh 1.000 g of the finely ground sample (Note 109) into the flask, add encapsulated stirring bar, and immediately attach a water-cooled condenser (with a standard 24/40 glass joint). Boil the solution in the flask on a magnetic stirrer hot plate for 20 min with mild stirring.

27.3.2.2 Remove the condenser and filter the contents of the flask on a small polypropylene Büchner funnel under vacuum, using a 250-mL filtering flask with side tube. Bring the filtrate to a boil and immediately titrate with standard ammonium acetate solution to a colorless end point.

27.3.3 *Calculation*—Calculate the percent free CaO to the nearest 0.1 % as follows:

$$\text{free CaO, \%} = EV \times 100$$

where:

E = CaO equivalent of the ammonium acetate solution, g/mL, and
V = millilitres of ammonium acetate solution required by the sample.

REFERENCES

(1) Crow, R. F., and Connolly, J. D., "Atomic Absorption Analysis of Portland Cement and Raw Mix Using Lithium Metaborate Fusion," *Journal of Testing and Evaluation,* Vol 1, No. 5, September 1973, pp. 382–393.

(2) Moore, C. W., "Suggested Method for Spectrochemical Analysis of Portland Cement by Fusion with Lithium Tetraborite Using an X-Ray Spectrometer," E-2 SM 10–26 in *Methods for Emission Spectrochemical Analysis,* ASTM, 1971.

(3) Jugovic, Z. T., "Applications of Spectrophotometric and EDTA Methods for Rapid Analysis of

Cement and Raw Materials," *Analytical Techniques for Hydraulic Cement and Concrete, ASTM STP 395*, ASTM, 1966, pp. 65–93.
(4) Bean, B. L., and Arni, H. T., "A New Rapid Method for Cement Analysis," (Atomic Absorption Spectrophotometry), Report No. FHWA-RD-72-41, Department of Transportation, Federal Highway Administration, September 1972. (Order copies from National Technical Information Service, Springfield, Va. 22151, by Order No. PB 243622.) Also see Sections 121 through 127 of AASHTO T105-73, Standard Specifications for Transportation Materials and Methods of Sampling and Testing, 11th ed., 1974.
(5) Bean, B. L., "Improvements in the Rapid Analysis of Portland Cement by Atomic Absorption Spectrophotometry," Report No. FHWA-RD-73-4, Department of Transportation, Federal Highway Administration, March 1973. (Order copies from National Technical Information Service, Springfield, Va. 22151, by Order No. PB-220-549.)

TABLE 1 Maximum Permissible Variations in Results[A]

(Column 1) Component	(Column 2) Maximum Difference Between Duplicates[D]	(Column 3) Maximum Difference of the Average of Duplicates from SRM Certificate Values[B,C,D]
SiO_2 (silicon dioxide)	0.16	±0.2
Al_2O_3 (aluminum oxide)	0.20	±0.2
Fe_2O_3 (ferric oxide)	0.10	±0.10
CaO (calcium oxide)	0.20	±0.3
MgO (magnesium oxide)	0.16	±0.2
SO_3 (sulfur trioxide)	0.10	±0.1
LOI (loss on ignition)	0.10	±0.10
Na_2O (sodium oxide)	0.03	±0.05
K_2O (potassium oxide)	0.03	±0.05
TiO_2 (titanium dioxide)	0.02	±0.03
P_2O_5 (phosphorus pentoxide)	0.03	±0.03
ZnO (zinc oxide)	0.03	±0.03
Mn_2O_3 (manganic oxide)	0.03	±0.03
S (sulfide sulfur)	0.01	F
Cl (chloride)	0.02	F
IR (insoluble residue)	0.10	F
Cx (free calcium oxide)	0.20	F
Alk_{sol} (water-soluble alkali)[E]	$0.75/w$	F
Chl_{sol} (chloroform-soluble organic substances)	0.004	F

[A] When all seven SRM cements are required, as for demonstrating performance of rapid methods, at least six of the seven shall be within the prescribed limits and the seventh shall differ by no more than twice that value. When a lesser number of SRM cements are required, all of the values shall be within the prescribed limits.

[B] Interelement corrections may be used for any oxide standardization provided improved accuracy can be demonstrated when the correction is applied to all seven SRM cements.

[C] Where an SRM certificate value includes a subscript number, that subscript number shall be treated as a valid significant figure.

[D] Where no value appears in Column 3, SRM certificate values do not exist. In such cases, only the requirement for differences between duplicates shall apply.

[E] w = weight, in grams, of samples used for the test.

[F] Not applicable. No certificate value given.

APPENDIX

(Nonmandatory Information)

X1. EXAMPLE OF DETERMINATION OF EQUIVALENCE POINT FOR THE CHLORIDE DETERMINATION

(Column 1) AgNO$_3$, mL	(Column 2) Potential, mV	(Column 3) Δ mV[A]	(Column 4) Δ^2 mV[B]
1.60	125.3		
1.80	119.5	5.8	1.4
2.00	112.3	7.2	1.3
2.20	103.8	8.5	1.3
2.40	94.0	9.8	0.6
2.60	84.8	9.2	2.3
2.80	77.9	6.9	0.8
3.00	71.8	6.1	1.3
3.20	67.0	4.8	

The equivalence point is in the maximum Δ mV interval (Column 3) and thus between 2.20 and 2.40 mL. The exact equivalence point in this 0.20 increment is calculated from the Δ^2 mV (Column 4) data as follows:

$$E = 2.20 + \frac{1.3}{1.3 + 0.6} \times 0.20 = 2.337 \text{ mL. Round to } 2.34.$$

[A] Differences between successive readings in Column 2.
[B] Differences between successive Δ readings in Column 3 "second differentials".

For additional useful information on details of cement test methods, reference may be made to the "Manual of Cement Testing," which appears in Vol 04.01 of the *Annual Book of ASTM Standards.*

The American Society for Testing and Materials takes no position respecting the validity of any patent rights asserted in connection with any item mentioned in this standard. Users of this standard are expressly advised that determination of the validity of any such patent rights, and the risk of infringement of such rights, are entirely their own responsibility.

This standard is subject to revision at any time by the responsible technical committee and must be reviewed every five years and if not revised, either reapproved or withdrawn. Your comments are invited either for revision of this standard or for additional standards and should be addressed to ASTM Headquarters. Your comments will receive careful consideration at a meeting of the responsible technical committee, which you may attend. If you feel that your comments have not received a fair hearing you should make your views known to the ASTM Committee on Standards, 1916 Race St., Philadelphia, PA 19103.